I0034489

Sustainable Digital Technologies for Smart Cities

This book focuses on recent and emerging techniques for the enhancement of smart healthcare, smart communication, and smart transportation systems. It covers topics ranging from Machine Learning techniques, the Internet of Things (IoT), security aspects of medical documents, the performance of various protocols used in the communication and transportation environment, simulation of systems for real-time applications, and overall analysis of the previously mentioned. Applications such as transportation systems, stock market prediction, Smart Cities, and vehicular communication are dealt with.

Features:

- Covers three important aspects of smart cities i.e., healthcare, smart communication and information, and smart transportation technologies.
- Discusses various security aspects of medical documents and the data preserving mechanisms.
- Provides better solutions using IoT techniques for healthcare, transportation, and communication systems.
- Includes the implementation example, various datasets, experimental results, and simulation procedures.
- Offers solutions for various disease prediction systems with intelligent techniques.

This book is aimed at researchers and graduate students in computer science, electrical engineering, and data analytics.

Sustainable Digital Technologies for Smart Cities

Healthcare, Communication, and Transportation

Edited by
L Ashok Kumar, R. Manivel
and Eyal Ben Dor

CRC Press
Taylor & Francis Group
Boca Raton London New York

CRC Press is an imprint of the
Taylor & Francis Group, an **informa** business

First edition published 2024
by CRC Press
6000 Broken Sound Parkway NW, Suite 300, Boca Raton, FL 33487-2742

and by CRC Press
4 Park Square, Milton Park, Abingdon, Oxon, OX14 4RN

CRC Press is an imprint of Taylor & Francis Group, LLC

© 2024 selection and editorial matter, L Ashok Kumar, R. Manivel and Eyal Ben Dor; individual chapters, the contributors

ISBN: 9781032309842 (hbk)
ISBN: 9781032310312 (pbk)
ISBN: 9781003307716 (ebk)

DOI: 10.1201/9781003307716

Typeset in Times
by Deanta Global Publishing Services, Chennai, India

Contents

About the Editors

L Ashok Kumar was a Postdoctoral Research Fellow from San Diego State University, California. He was selected among seven scientists in India for the BHAVAN Fellowship from the Indo–US Science and Technology Forum and also received the SYST Fellowship from DST, Government of India. He has 3 years of industrial experience and 22 years of academic and research experience. He has published 173 technical papers in international and national journals and has presented 167 papers at national and international conferences. He has completed 26 Government of India–funded projects worth about 150,000,000 rupees and currently he has 9 projects in progress worth about 120,000,000 rupees. He has developed 27 products and out of those 23 products have been technology transferred to industries and for government funding agencies. He has created 8 Centres of Excellence at PSG Tech in collaboration with government agencies and industries namely, the Centre for Audio Visual Speech Recognition, the Centre for Alternate Cooling Technologies, the Centre for Industrial Cyber Physical Systems Research, the Centre for Excellence in LV Switchgear, the Centre for Renewable Energy Systems, the Centre for Excellence in Solar PV Systems and the Centre for Excellence in Solar Thermal Systems. His PhD work on wearable electronics earned him a National Award from the ISTE, and he has received 26 awards at both the national and international levels. He has guided 92 graduate and postgraduate projects. He has produced 6 PhD scholars, and 12 candidates are doing PhDs under his supervision. He has visited many countries for institute industry collaboration and as a keynote speaker. He has been an invited speaker in 345 programs. Also, he has organized 102 events, including conferences, workshops, and seminars. He completed his graduate program in electrical and electronics engineering at the University of Madras, India, postgraduate from PSG College of Technology, India, and master's in business administration from IGNOU, New Delhi, India. After completing his graduate degree, he joined Serval Paper Boards Ltd., Coimbatore (now ITC Unit, Kovai) as a project engineer. Presently he is working as a Professor in the Department of EEE, PSG College of Technology, India. He is also a Certified Charted Engineer and BSI Certified ISO 500001 2008 Lead Auditor. He has authored 19 books in his areas of interest with various publishers. He has 11 patents, 1 design patent and 2 copyrights to his credit, and has also contributed 18 chapters in various books. He is also the Chairman of the Indian Association of Energy Management Professionals; an Executive Member of the Institution of Engineers (IE), Coimbatore; an Executive Council Member of the Institute of Smart Structure and Systems, Bangalore, and an Associate Member of CODISSIA. He also holds prestigious positions in various national and international forums and is a Fellow Member of the IET (UK), a Fellow Member of the IETE, a Fellow Member of the IE and a Senior Member of the IEEE.

R. Manivel is a Professor of Mechanical Engineering and Head of Research at Kumara Guru College of Technology, Coimbatore, India. Research areas include thermal energy storage and computational fluid dynamics with fluid surface interaction. In industries he carried out industrial design analysis and experimental techniques in the thermal equipment applications of engines, blowers, compressors, fans, and heat exchangers. He has received research grants of 9,000,000 rupees from Government funding agencies to carry out research projects in the area of energy storage, automobile radiators, hydro power generation, etc. He has published 50 scientific papers in various proceedings and journals. He is a Fellow Member of the Institution of Engineers (IE) India and a Member of the Combustion Institute, India. The areas of his academic interest include energy engineering, modeling, computational fluid dynamics, and energy transfer and storage. Further, in the field of oil and gas industries, his academic interests are subsea valves and equipment design through ASME standards.

Eyal Ben Dor is the Head of the Remote Sensing Laboratory at Tel Aviv University (RSL-TAU), Israel, the leading group in Israel for imaging spectroscopy (IS), soil spectroscopy, and remote sensing applications dedicated to soil mapping and environmental monitoring. The RSL-TAU is an expert in the field and airborne campaigns at national and international levels and has gained significant experience to that end in many projects. IS technology is also used by the RSL-TAU to investigate issues in medicine, veterinary, food security, civil engineering, and agriculture. The RSL-TAU is responsible for more than 200 scientific papers in scientific proceedings, peer review journals, and book chapters. Four patents in IS for veterinary, civil engineering, soil science, and contamination application also belong to the RSL-TAU. The areas of Ben Dor's academic interest include hyperspectral remote sensing of the earth, soil spectroscopy, developing applications in many disciplines (medicine, veterinary, civil engineering, precise agriculture), generating a soil spectral library for Mediterranean countries, and bilateral projects with the Czech Republic (CGS), Germany (GFZ), Greece (i-BEC), Italy (CNR), and EC (EUFAR, EO-MINERS, and GEO-CRADEL).

Contributors

N. Akshaya has completed their Bachelor of Engineering in computer science and engineering from the Kumaraguru College of Technology, Coimbatore, India

V. Amarnath graduated from the Department of Electronics and Communication Engineering, PSG College of Technology, Coimbatore, India, in the year 2020 and completed his postgraduation (MTech) in Communication Engineering from Vellore Institute of Technology, Vellore, India in the year 2022. He worked as a software engineer in L&T Technology Service, Mumbai, India, for a brief period of 5 months before taking up the role of Director-GIS in Skyx Aerospace Ltd. His current research area includes rainfall and drought analysis and prediction using drone-enabled solutions.

A. Amritha has completed their Bachelor of Engineering at Kumaraguru College of Technology in the Department of Computer Science and Engineering, Coimbatore, India.

Kumara Krishna Ananthan Viveka Vikram is an undergraduate scholar in the Department of Electronics and Communication Engineering, Kumarguru College of Technology, Coimbatore, India. Their areas of interest include Artificial Intelligence and signal processing.

Tamilselvan Arunachalam has completed their Bachelor of Engineering in instrumentation and control engineering from the PSG College of Technology, Coimbatore, India.

R. Atshayasri has completed their Bachelor of Engineering at Kumaraguru College of Technology in the Department of Computer Science and Engineering, Coimbatore, India.

K. G. Sethu Balaji has completed their Bachelor of Engineering in computer science and engineering from the Kumaraguru College of Technology, Coimbatore, India.

K. R. Baskaran pursued his Bachelor of Engineering in computer engineering from Madurai Kamaraj University, Madurai, India, Master of Science in software systems from the Birla Institute of Technology and Science, Pilani, India, Master of Engineering in computer science and engineering from Kumaraguru College of Technology, Coimbatore, India, and obtained his PhD from Anna University, Chennai, India, in 2015. His areas of interest are data analytics, Machine Learning, operating systems, and compiler design. He has 32 years of teaching experience and is currently working as a professor in the Department of Computer Science and Engineering at the Kumaraguru College of Technology, Coimbatore, India. Presently, he is guiding six research scholars affiliated with Anna University, Chennai, India.

D. Jagadeesan works as an assistant professor in the Department of Commerce PA and IT at Kaamadhenu Arts and Science College, Sathyamangalam, India. He received his Bachelor of Science in electronics at SNR Sons College, Bharathiar University, Coimbatore, India, in the year 1994. He completed his Master of Business Administration at RVS College, Bharathiar University, Coimbatore, India, in the year 2000. He completed his Master of Philosophy in Management at SNR Sons College, Bharathiar University, Coimbatore, India, in the year 2011. He has published many papers in international journals and conferences.

Umashankar Jugal Balaji is currently working as an associate design engineer at Bosch Global Software Technologies, Coimbatore, India. He received his bachelor's degree in mechanical engineering from the PSG College of Technology, Coimbatore, India, in 2019 following which he joined Bosch Global Software Technologies.

J. Cynthia is currently working as a professor in the Department of Computer Science and Engineering, at the Kumaraguru College of Technology, Coimbatore, India. She has completed her Bachelor of Engineering and Master of Engineering in computer science and engineering and has received a PhD from Anna University, Chennai, India. She has over 22 years of experience in industry and academia. She has over 25 international journal articles, 5 book chapters, and 15 international conference publications to her credit. She has completed over 15 certification courses from various international universities. She is a member of the ISTE and the IET. She is a recipient of various international awards. She has completed five consultancy projects for various industries and government organizations. Her areas of interest include wired and wireless network protocols, vehicular ad hoc networks (VANETs), the Internet of Things (IoT), blockchain, and smart city applications.

Sivaraj Dhakshin is an undergraduate scholar in the Department of Electronics and Communication Engineering, Kumarguru College of Technology, Coimbatore, India. Their areas of interest include Artificial Intelligence and signal processing.

S. J. Syed Ali Fathima was working as an assistant professor at the Kumaraguru College of Technology, Coimbatore, India, and has specialized in Artificial Intelligence, Machine Learning, and augmented/virtual reality.

Deenadayalan Karaiyagowder Govindarajan is a postgraduate student in the Department of Biotechnology at Kumaraguru College of Technology, Coimbatore, India.

Rajaguru Harikumar received his Bachelor of Engineering (ECE) from the National Institute of Technology (REC), Trichy, India, in the year 1988. He obtained his Master of Engineering in applied electronics from the College of Engineering, Anna University, Chennai, India, in 1990. He was awarded his PhD in information and communication engineering from Anna University, Chennai, India, in 2009. He has 32 years of teaching experience at college level. He worked as faculty at the

Department of ECE, PSNA College of Engineering and Technology, Dindigul, India. He was an assistant professor in the IT Department at PSG College of Technology, Coimbatore, India. He also worked as an assistant professor in the ECE Department at Amrita Institute of Technology, Coimbatore, India. Currently, he is working as a professor of ECE at Bannari Amman Institute of Technology, Sathyamangalam, India. He has published 258 papers in international and national journals and also published around 316 papers in international and national conference proceedings conducted both in India and abroad. He has completed four sponsored projects with a value of INR 2,600,000 from agencies such as the AICTE, the DRDO, and the ICMR. He is the recipient of the IETE RS Khandpur Award 2022 for his contribution toward medical electronics instrumentation. His areas of interest are biosignal processing, soft computing, medical image processing, VLSI design, and communication engineering. Currently, he is guiding 10 PhD candidates in these areas. Eighteen of his students were awarded PhDs by Anna University, Chennai, India. He is a lifetime member of the IETE, the IAENG, and the ISTE and a member of the IEEE.

Thamizhi Shanmugam Indrani has completed her bachelor's in computer science and engineering at the PSG College of Technology, Coimbatore, India.

Kumaraswamy Jasmine is an assistant professor in the Department of Electronics and Communication Engineering, at the Kumaraguru College of Technology, Coimbatore, India. She received her Master of Engineering in communication systems from Anna University, Coimbatore, India, and completed her Bachelor of Engineering in electronics and communication engineering from Manonmaniam Sundaranar University, Thirunelveli, India. She is currently conducting research in the field of signal processing for wireless communication at Anna University, Chennai, India. She has more than 15 years of experience in teaching and has published papers in both international journals and conferences. Her research interests include machine-to-machine communication and spectrum sensing.

R. Kalaiselvi is currently working as an assistant professor at Kumaraguru College of Technology in the Department of Information Science and Engineering, Coimbatore, India. She received her PhD in cloud computing from Anna University, Chennai, India. Her research focuses primarily on cloud security. She has 20 years of teaching experience. She has around 20 publications in various journals and conference proceedings. Her research interests include cloud computing, cloud security, and blockchain.

G. R. Karpagam is a professor with 26 years of experience in the Department of Computer Science and Engineering at PSG College of Technology. Coimbatore, India. She holds the additional responsibility of professor in charge of the Dr. GRD Memorial Library. Karpagam received her bachelor's degree, master's degree, and PhD in computer science and engineering. Her research interest is in areas related to Artificial Intelligence (AI), Machine Learning (ML), service-oriented architecture, model-driven architecture, security, and cloud computing. She is a recipient of the Chartered Management Institute's (CMI's) Level 5 award (management and leadership) and the PSG & Sons Teacher of the Year Award. She has been conferred with senior membership in the IEEE, Senior IEEE – Women in Engineering, Consultant and Computer Society. She also holds professional membership in the Association for Computing Machinery (ACM), the Indian Society for Technical Education, and the Indian Institution of Engineers.

Lekha Kaushik is an undergraduate student in the Department of Biotechnology at Kumaraguru College of Technology, Coimbatore, India.

Kanakaraj Kavitha received her Bachelor of Engineering in electronics and communication engineering at the Institute of Road and Transport Technology, Erode, India, her Master of Engineering in communication systems at the PSG College of Technology, Coimbatore, India, and her PhD from Anna University, Chennai, India. She works at the Kumaraguru College of Technology, Coimbatore, India, as a professor in the Department of Electronics and Communication Engineering. She has 25 years of teaching and research experience. She has published 3 book chapters, 28 research papers in national and international journals, and 60 papers at conferences. Her research interests include signal processing, wireless communication systems, and Machine Learning techniques. Her research projects have been funded by the AICTE under the RPS scheme, the DRDO, and KCT management.

B. Kaviya has completed their Master of Engineering in computer science and engineering from the Kumaraguru College of Technology, Pilani, India.

D. Sathya Associate Professor, School of Computer Science and Engineering, RV University, Bengaluru, India

Shanmugam Arun Kumar is working as an assistant professor at Kumaraguru College of Technology, Coimbatore, India. Currently, he is pursuing a PhD at Anna University, Chennai, India, in computational intelligence in signal and image processing. He completed his bachelor's degree in electronics and communication engineering from Kumaraguru College of Technology, Coimbatore, India. Further, he completed his Master of Technology in communication engineering and signal Processing from Amrita University, Coimbatore, India. He is an active member of the IEEE and IE(I) professional society. He is an avid researcher in signal processing, image processing, biomedical systems, and Machine Learning. To his credit, he has around 35 technical papers published in international peer-reviewed and Scopus-indexed journals and conferences.

Shanmugham Ashwin Kumar is currently working as a business analyst in the Banking & Capital Markets Vertical for Genpact, Bengaluru, India. He received his bachelor's degree in Mechanical Engineering from the PSG College of Technology, Coimbatore, India in 2019. Following this, he worked on an engine assembly line and was in charge of the green field project for Hero Motocorp between 2019 and 2021.

After that, he did a course in data analytics from the IIT Kolkata, Kolkata, India, before joining Genpact.

N. Vinoth Kumar completed his PhD in electrical engineering from the Central Power Research Institute, Bangalore, India, in 2018. He has over 17 years of experience in teaching, research, and the electrical industry. He was looking after many power plant/substation projects for various agencies such as NHPC, PGCL, and KPTCL. His research areas focus on inverters for microgrid applications, smart grids, and electric vehicles. He has a strong interest in the development of multilevel inverters for various levels and the optimization algorithms for the reduction of THD. His current focus is on electric vehicles, fuel-based vehicles, and power converters for various e-mobility applications. He has published many papers in reputed journals and conferences in various international forums.

Kandaswamy Kumaravel is an assistant professor at the Department of Biotechnology, Kumaraguru College of Technology, Coimbatore, India. He received his PhD in molecular microbiology and biophysics from Nanyang Technological University (NTU), Singapore. His research focuses primarily on developing novel antimicrobial compounds to combat microbial infections. Furthermore, he has a strong interest in developing high-throughput screening platforms to identify novel drug candidates. He is currently involved in developing state-of-the-art high-resolution imaging platforms to quantify the dynamics of subcellular molecules in bacterial cells and biofilms. He is a principal investigator (PI) on the SERB Startup Research Grant (~INR 3,300,000) awarded by the Department of Science and Technology, Government of India. He is also a co-recipient of the FoldScope grant (BT/IN/INDO-US/Foldscope/39/2015) awarded by the Department of Biotechnology, Ministry of Science and Technology, Government of India.

L. Latha is currently working as a professor in the Department of Computer Science and Engineering at the Kumaraguru College of Technology, Coimbatore, India. She has over 25 years in academia and industry. She completed her doctorate, in the field of "biometric authentication", at Anna University, Chennai, India. She won the "Best Ph.D. Thesis Award" from the Computer Society of India and achieved second place at the national level. Her research interests include multimodal biometrics, network security, pattern recognition, and digital image processing. She has published many papers in journals and various conference proceedings and she has won the "best paper award" three times. She has completed a funded research project in the area of biometric access control and holds a research guideship at Anna University, Chennai, India.

Praveen Kumar Mageswaran completed their Bachelor of Engineering in instrumentation and control engineering from the PSG College of Technology, Coimbatore, India.

Krishnamoorthi Maheswari is an astute professional with 18 years of teaching experience. She has completed her Master of Science and Master of Philosophy in mathematics at Bharathiar University, Coimbatore, India. She has also completed her doctorate from Bharathiar University, Coimbatore, India. Her areas of interest include stability analysis, neural networks, and non-linear dynamics. She has published articles in SCI-indexed journals. To her credit, she has 251 citations with h-index 5 and i-10 index 4. She is a life member of the Indian Mathematical Society, the Ramanujan Mathematical Society, and also an IEEE member. Currently, she is heading the Department of Mathematics at Kumaraguru College of Technology, Coimbatore, India. She has acted as a session chair and resource person in various international conferences, and also as a reviewer in the SCI, SCIE, WOS, and Scopus-indexed journals. Department of Mathematics, China, Research Center for Wind Energy Systems, Kunsan National University, Gunsan, South Korea, School of IT Technology and Control Engineering, Kunsan National University, South Korea, School of Mathematical Sciences, Chongqing Normal University, Chongqing, People's Republic of China, Department of Electrical Engineering, Yeungnam University, Kyongsan, South Korea. She has collaborated with various scientists across the globe and has been to various international conferences for presentations such as at the National University of Singapore, the University of Bristol, UK, the University of Bath, UK, and the University of Oxford, UK.

S. Uma Maheswari was born in Coonoor, India, in 1989. She completed her Ph.D in the year 2021 and received her B.Tech degree in Information Technology from the Anna University, Chennai, India, in 2010, and the M.E degree in Computer Science and Engineering from Anna University, Chennai, India, in 2012.

Since 2012, she has been with the Department of Computer Science and Engineering, Kumarguru College of Technology as Assistant Professor. Her current research interests include Multimedia Security, Health Care Informatics, Cloud Computing and image processing techniques.

Yogesan Meganathan is a postgraduate student in the Department of Biotechnology at Kumaraguru College of Technology, Coimbatore, India.

Santhakumar Mohan is an associate professor at the Department of Mechanical Engineering and Dean of Industry Collaboration and Sponsored Research at the Indian Institute of Technology (IIT), Palakkad, India. He received his PhD in robotics and control from the Indian Institute of Technology Madras, Chennai, India, in 2010. From June 2010 to March 2011, he worked as an assistant professor in the Department of Mechanical Engineering at the National Institute of Technology Calicut (NITC), Kerala, India. He then worked as a postdoctoral fellow at the Korean Advanced Institute of Science and Technology (KAIST), Daejeon, Republic of Korea. In 2012, he joined the faculty of Mechanical Engineering at the Indian Institute of Technology, Indore, India. He holds visiting faculty positions at IISc Bangalore, India; BSTU, Belgorod, Russia; KAIST, Daejeon, Republic of Korea; RWTH Aachen, Germany; and Épreuves Classantes Nationales (ECN), France. His active research areas include service and field robots specifically, the design and motion control of underwater and wheeled mobile robots, parallel robotic platforms, and assistive and rehabilitation robots. Furthermore, he received the outstanding young scientist award for the year 2014 from the Korea Robotics Society, the European Master on Advanced Robotics Plus (EMARO+) fellowship (2018–2019), the Alexander von Humboldt (AvH) fellowship (2016–2017), the Satellite Across Virtual Institutes (SAVI) fellowship (2013–2014), the world-class university fellowship (2011–2012), and the Brain Korean 21 (BK2) fellowship (2011). He has published more than 120 articles in various journals and conference proceedings. He has edited three books and has completed nine sponsored projects and currently has nine ongoing sponsored projects.

Sowmiya Muthumanickam is an undergraduate student in the Department of Biotechnology at Kumaraguru College of Technology, Coimbatore, India.

S. P. Naveen has completed their Bachelor of Engineering in instrumentation and control engineering from the PSG College of Technology, Coimbatore, India.

V. Gokul Nitheesh has completed their Bachelor of Engineering in instrumentation and control engineering from the PSG College of Technology, Coimbatore, India.

Arun Kumar Pinagapani is an assistant professor, senior grade, in the ICE Department, at the PSG College of Technology, Coimbatore, India. He received his Bachelor of Technology in electronics and instrumentation engineering from SASTRA University, Thanjavur, India, and a Master of Engineering in control systems from PSG College of Technology, Coimbatore, India. He completed a PhD under the Faculty of Electrical Engineering, at Anna University, Chennai, India. His area of interest is advanced process control.

Krishnagoundenpudur Natarajan Anu Prabha is an undergraduate scholar in the Department of Electronics and Communication Engineering, Kumarguru College of Technology, Coimbatore, India. Their areas of interest include Artificial Intelligence and signal processing.

Ramasamy Dhivya Praba received her Bachelor of Engineering degree in electronics and communication engineering at the SSM College of Engineering, Namakkal, India, and her Master of Engineering in applied electronics at the Government College of Technology, Coimbatore, India. Currently, she works at Kumaraguru College of Technology, Coimbatore, India, as an assistant professor in the Department of Electronics and Communication Engineering. She has 14 years of teaching and research experience. She has published 1 book chapter, 16 research papers in national and international journals, and 20 papers at conferences. Her research interests include signal/image processing, Machine Learning techniques, and embedded systems.

K. Pranavi completed her Bachelor of Engineering in computer science and engineering at Kumaraguru College of Technology, Coimbatore, India, in the years

2014–2018 and is undergoing her Master of Engineering in computer science and engineering at Kumaraguru College of Technology, Coimbatore, India.

Selvaraj Rajini has 25 years of collegiate teaching experience. She has a doctorate from Anna University, Chennai, India, specializing in "word sense disambiguation techniques" using Artificial Intelligence and Deep Learning. She has organized several intercollegiate technical competitions and training programs and has been invited to deliver talks on the latest developments in computer science and engineering.

Soundarya Ravichandran has completed her bachelor's in computer science and engineering at the PSG College of Technology, Coimbatore, India.

G. Sakthipriya is currently working as an assistant professor at the Department of Computer Science and Engineering, Ramco Institute of Technology, Rajapalayam, India. She has completed her postgraduate program, a Master of Engineering in computer science and engineering from the Kumaraguru College of Technology, Coimbatore, India. Her research focuses on VANET. She has worked in real-time traffic simulation using network simulators. She has been awarded "Best Project" for the academic year 2019–2020. She has a strong interest in Machine Learning and is working on projects such as time series forecasting and deep neural networks. She is currently working in 3D motion analysis for enhancing augmented reality and virtual reality.

S. Sangeetha is currently working in the Tamil Nadu State Council for Science and Technology (TNSCST), she has 13 years of teaching experience and 5 years of research experience. She has published more than 30 papers in reputed journals and conferences. She has had two projects funded by the TNSCST and Xambala Inc. She has also organized various workshops, seminars, and conferences funded by the Defence Research and Development Organisation (DRDO), the AICTE, and the TNSCST. She is a review member and editorial board member in many international conferences and journals.

Dinakari Sarangan is an undergraduate student in the Department of Biotechnology at Kumaraguru College of Technology, Coimbatore, India.

N. Saritakumar is an assistant professor in the Department of Electronics and Communication Engineering, PSG College of Technology, Coimbatore, India. He has 18 years of experience in the telecommunication industry and academics. His research interests include wireless networks and software-defined networks.

Shanmugam Sasikala received her Bachelor of Engineering (ECE) from Bharathidasan University, Trichy, India, in 1998. She obtained her Master of Technology (biomedical engineering) from the Indian Institute of Technology, Madras, India, in 2005. She has 22 years of teaching experience at the college level. She completed her Doctoral program in information and communication engineering in 2018 at Anna University, Chennai, India. Currently, she is working as an associate professor in the Department of ECE at Kumaraguru College of Technology, Coimbatore, India. She is an active member of the IEEE, the ISTE and the IE(I). Her areas of interest include signal processing, image processing, biomedical systems, and Machine Learning. To her credit, she has around 60 technical papers published in international peer-reviewed Scopus and SCI-indexed journals and conferences.

S. N. Shivappriya is highly innovative and hard-working towards finding solutions for complex real-work problems. As a facilitator, she has also motivated students to build careers in their areas of interest. She received her Bachelor of Engineering (ECE) from Periyar University, Salem, India, in 2004. She obtained her Master of Engineering degree in communication systems from Anna University, Chennai, India, in 2006. She completed her Ph.D. in bio-medical signal processing from Anna University, Chennai, India, in 2015 and has 16 years of teaching experience at college level. Currently, she is working as an associate professor in the Department of ECE at Kumaraguru College of Technology, Coimbatore, India. Her areas of interest include signal processing, image processing, soft computing and embedded communication systems. She is an active member of the IEEE and IEI. She has 30 publications in international and national journals, 4 book chapter publications and 70 publications in international and national conference proceedings. She has received 10 grants from the CSIR, the IEI, the IEEE, the TNSCST and from industry for conducting seminars, conferences and workshops. She received a TNSCST travel grant for attending an international conference in Japan in 2019.

Ms. Amrithaa Sivakumar graduated with a Bachelors in Engineering from the Department of Electronics and Communication Engineering, PSG College of Technology in the year 2020. Currently working as Application analyst at Citi. She is interested in recent technologies and has worked on projects involving LoRa and Machine Learning, with a view of using these technologies for improved and advanced Agricultural processes.

Muthusaravanan Sivaramakrishnan is a PhD Student at Chemical Glycobiology Lab, BITS-pilani. He holds M.tech & B.Tech in Biotechnology. His research focuses primarily on Nature-Inspired Optimization strategies, development of modified biomaterials for Bioremediation and antimicrobial applications.

Dhana Srinithi Srinivasan completed her bachelor's in Computer Science and Engineering at PSG College of Technology, Coimbatore, India, and is currently pursuing a Master of Data Science at the Johns Hopkins University, MA. She is keen on exploring various opportunities and technologies in the field of research in Computer Science, with a passion for data. Her eventual goal is to develop tools and technologies for the benefit of society.

Adarsh V Srinivasan graduated from the Department of Electronics and Communication Engineering, the PSG College of Technology, Coimbatore, India, in the year 2020. He graduated with the award of the Best Outgoing Student for his class and worked for Cisco Systems from January 2020 to July 2020 as an Intern and as a full-time employee from August 2020 to October 2021. He then started his own startup "Skyx Aerospace Ltd." in March 2022 which focuses on manufacturing unmanned aerial vehicles (Drones) for various applications (agriculture, survey and defense) and for solving many problems using drone-enabled solutions. He and his firm have currently partnered with multiple government organizations and other multinational corporations (MNCs) to benefit people in various ways apart from this, he is conducting research on the topic of "drones in precision agriculture".

V. Sudha was working as an assistant professor at Kumaraguru College of Technology in the Department of Computer Science and Engineering, Coimbatore, India. She received her PhD from Anna University, Chennai, India. She has more than 10 years of teaching experience.

Jayaseelan Clement Sudhahar is currently working as a professor and is the Head of the Department of Management Studies, at Karunya Deemed University, Coimbatore, India. He has over 30 years of experience. His first decade of experience was in the marketing industry working for companies such as M&M. His next two decades were in the education industry as faculty and an administrator. He involved himself in research and consultancy. He has completed two applied research projects funded by the Indian Council of Social Science Research (ICSSR) for the elevation of the Micro, Small & Medium Enterprises (MSME) sector. He has penned over 10 books and published over 80 articles with more than 1,000 citations to his credit. He has also overseen 14 PhD candidates in his main area of interest, digital branding, and marketing analytics.

N. Suganthi is a professor with 22 years of teaching experience, as well as over 40 conference and research journal publications. She completed her PhD in 2014 and received Anna University's, Coimbatore, India, supervisor recognition in 2016. She oversees courses in hardware, security, and wireless networks. She has guided two to three undergraduate and postgraduate projects each of her teaching years and is currently guiding five PhD scholars, including one full-time scholar. She spends her free time reading books and gardening.

M. Suguna is an assistant professor, senior grade II in the School of Computer Science and Engineering at the Vellore Institute of Technology, Chennai, India. Her research interests include data analytics, health care analytics, cloud computing, and agile project management. Currently, she has completed a PhD in information and communication engineering from Anna University, Chennai, India, an undergraduate Bachelor of Engineering in computer science and engineering at Kumaraguru College of Technology, Coimbatore, India, and a Master of Engineering in computer science and engineering at Government College of Technology, Coimbatore, India. She has published more than 20 research articles in international journals and has been published in the proceedings of 15 international conferences. She is a member of the ISTE and the International Association of Engineers (IAENG).

V. P. Sumathi is currently working as an associate professor at Kumaraguru College of Technology in the Department of Computer Science and Engineering, Coimbatore, India. She received her PhD in data analytics from Anna University, Chennai, India. Her research focuses primarily on IoT and the application of Machine Learning algorithms in industry. She has 20 years of teaching experience. She has around 30 publications in various journals and conference proceedings. Her research interests include semantic web, data mining, and big data analytics.

Kandhasamy Thilagavathi has more than 18 years of teaching experience at both undergraduate and postgraduate levels. Currently, she is working as an assistant professor in the Department of Electronics and Communication Engineering, at the Kumaraguru College of Technology, Coimbatore, India. She has published technical papers in various international journals and conferences. Her areas of interest include hyperspectral remote sensing, communication systems, and microwaves.

Sadhasivam Udhaykumar is an assistant professor at the Department of Mechanical Engineering, PSG College of Technology, Coimbatore, India. His areas of interest include manufacturing, mechatronics, robotics, and automation. His research focuses primarily on manufacturing automation and energy harvesting. He has been involved in two technology transfers. He has one patent granted and three patents published. He has worked with research projects sponsored by DST, All India Council for Technical Education (AICTE), and UGC. He has conducted several training programs for industries such as Ashok Leyland, RANE, TAFE, L&T e-Engg solutions, CUMI, Hansen Drives, DRDL, NALCO, etc. He was awarded the Young Achiever Award in 2019 and the Young Engineer Award 2016–17 both from the Institution of Engineers, India He also received the PSG Teacher of the Year award in 2017. He has delivered more than 60 guest lectures to industrial/academic participants. He has organized more than 30 sponsored workshops/training programs for faculty members and students. He is a member of several professional bodies such as the Institution of Engineers (IE), the Indian Society for Technical Education (ISTE), the Tribology Society of India (TSI), and the Institute of Smart Structures and Systems (ISSS). He is a reviewer for several reputed international and national journals. He is currently involved in developing state-of-the-art high-resolution imaging platforms to quantify the dynamics of subcellular molecules in bacterial cells and biofilms. He is a Principal Investigator (PI) on the SERB Startup Research Grant (~INR 3,300,000) awarded by the Department of Science and Technology, Government of India. He is also a co-recipient of the FoldScope grant (BT/IN/INDO-US/Foldscope/39/2015) awarded by the Department of Biotechnology, Ministry of Science and Technology, Government of India.

C. Vasanthanayaki received a Bachelor of Engineering in electronics and communication engineering and a master's degree in computer science, both from the Government College of Engineering, Tirunelveli, India, in 1987 and 1997, respectively. She has pursued research on image analysis under the guidance of Dr. S. Annadurai at Anna University, Chennai, India. She has published more than 15 research papers in both national and international conference proceedings.

Prakash Vishnuvarthan is currently pursuing his master's degree in Business Administration at the Bharathidasan Institute of Management, Trichy, India. He received his bachelor's degree in Mechanical Engineering from the PSG College of Technology, Coimbatore, India, in 2019. Following this, he worked as an engineer in the Department of Maintenance at MRF, Trichy, India, between 2019 and 2021.

N. Yashwanth has completed their Bachelor of Engineering in computer science and engineering from the Kumaraguru College of Technology, Coimbatore, India.

Preface

This book focuses on recent and emerging techniques for the enhancement of smart healthcare, smart communication, and smart transportation system. It covers topics ranging from Machine Learning techniques, the Internet of Things, security aspects of medical documents, the performance of various protocols used in the communication and transportation environment, simulation of systems for real-time applications and their overall analysis. In addition, it provides good solutions to complex and unpredictable problems, thus helping the readers to understand more about new inventions in these areas. Also, the reader gains knowledge on developing products that are of social importance. In this book, applications such as transportation systems, stock market prediction, smart cities, and vehicular communication are dealt with. Smart communication provides a fast and efficient way to handle day-to-day challenges such as traffic management, waste management, security, energy management, etc. Machine Learning and IoT techniques are adopted in smart communication that helps to take a preventive decision and appropriate action whenever required. The book also discusses the performance of various protocols used in the communication environment.

FEATURES

- Discusses various security aspects of medical documents and data preserving mechanisms.
- Provides better solutions using IoT techniques for healthcare, transportation, and communication systems.
- Implementation examples, various datasets, experimental results, and simulation procedures.
- Solutions for various disease prediction systems with intelligent techniques.
- Students and researchers in the scientific community can get a better experience in the real-time implementation of various smart technologies to suit their specific needs.

1 Comparative Analysis of Selective Harmonic Elimination for Multilevel Inverters with CSA, PSO, BA, and SA

N. Vinoth Kumar

CONTENTS

1.1 INTRODUCTION

Multilevel inverters (MLIs) are used in medium and high power applications [1]. Multilevel inverters operate at a lower switching frequency in comparison with sine Pulse Width Modulated (PWM) inverters. Hence the losses are less in comparison with two-level inverters. The significant advantages of multilevel configuration are reduction in Total Harmonic Distortion (THD) in the output waveform without increasing switching frequency or decreasing the power output from an inverter. Due to the exclusion of transformers from coupling, a reduction in size and volume can be achieved. MLI has additional advantages: lower dv/dt ratio and less stress on power switches. Comparing two-level inverter topologies at the same power ratings, MLIs also have the advantage of line-to-line voltage harmonics being reduced owing to their switching frequencies. There are three traditional multilevel inverters categorized based on the clamping devices

DOI: 10.1201/9781003307716-1

and connection configuration [2]. They are diode-clamped multilevel inverters, capacitor-clamped multilevel inverters and cascaded multilevel inverters. Cascaded inverters are commonly used as they need fewer components in comparison with other traditional topologies. It is important that the output produced by the inverter is free from distortion. To produce quality output power from an inverter, many control and modulation schemes are being researched. A few are Sine Pulse Width Modulation (SPWM), Selective Harmonic Elimination (SHE) and space vector modulation techniques. In SHE, the lower-order harmonics can be removed by properly selecting switching angles. Different optimization algorithms such as Genetic Algorithm (GA), symmetric polynomial and resultant theory and bee algorithm have been used to suppress THD which reduces the lower harmonics. In this chapter, the four algorithms, particle swarm optimization, bat algorithm, simulated annealing and particle swarm optimization (PSO) have been compared for nine-level inverters. The code for the objective has been developed using MATLAB® and it is verified experimentally.

1.2 CASCADED H-BRIDGE MULTILEVEL INVERTER (CMLI)

The thought behind cascaded MLI is cascading each H-bridge output provides a stepped or staircase output. Each module of an H-bridge inverter consists of a Direct Current (DC) source and four switches. Each module is capable of producing positive V_{DC}, negative V_{DC} and zero voltage. Figure 1.1 shows a nine-level cascaded multilevel inverter.

To obtain higher voltage levels, additional modules can be added. In the case of nine-level inverters, four modules are required. Symmetric and asymmetric are the two categories of cascaded MLI.

1.2.1 FORMATION OF FITNESS FUNCTION

$$V(\omega t) = \begin{cases} 0 & n = \text{even} \\ \dfrac{4V}{n\pi} \sum_{i=1}^{s} \cos(n\alpha_i) & n = \text{odd} \end{cases} \tag{1.1}$$

Where "s" is the DC source count and "n" is harmonic order.

The waveform shown in Figure 1.2 is a pictorial representation of a nine-level waveform. The voltage sources (Vdc) are the same and the switching angle is selected from zero to $\pi/2$. Only odd harmonics appear in the output waveform whereas the even harmonics are zero due to symmetry in nature. The Fourier series expansion for the stepped voltage output waveform is given in Equation (1.1).

Equation (1.2) expresses the THD value—[3]. Vn is the amplitude of the nth harmonic which can be obtained using Equation (1.3), and $V1$ is the amplitude of fundamental voltage and α_k are the switching angles per quarter.

$$THD = \sqrt{\sum_{2}^{n} V_{n^2}} \Big/ V_1 \tag{1.2}$$

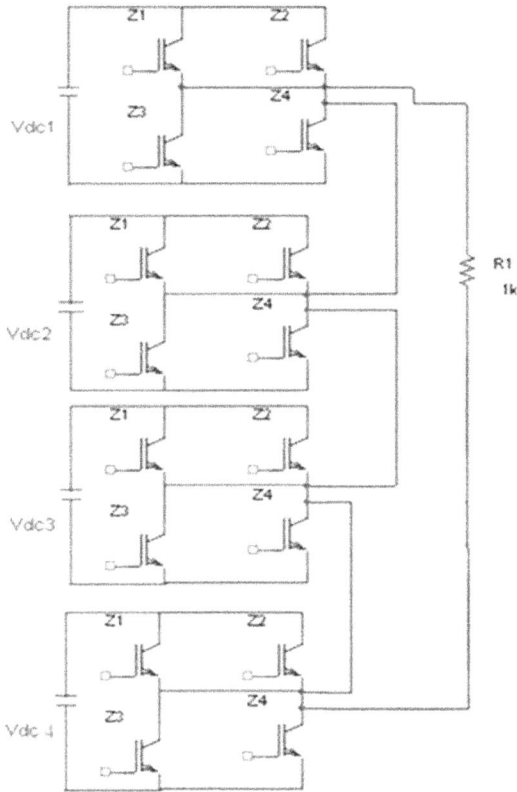

FIGURE 1.1   Nine-level cascaded MLI

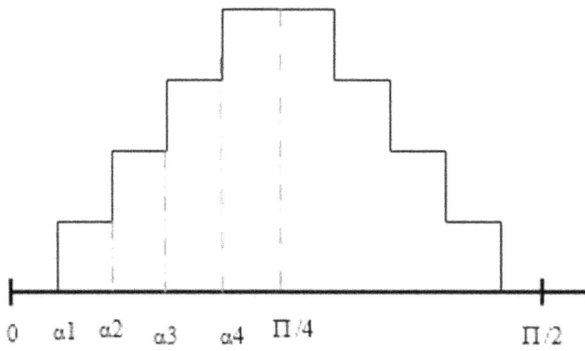

FIGURE 1.2   Step waveform for nine levels

$$V_n = \left(\left(\frac{4V}{\pi} \right) \sum_{2}^{n} \sum_{1}^{k} \frac{Cos(n\alpha_k)}{n} \right) \tag{1.3}$$

$$V_1 = \left(\left(\frac{4V}{\pi} \right) \sum_{2}^{k} Cos(n\alpha_k) \right) \tag{1.4}$$

The SHE technique aims to remove/suppress the selected harmonics which are usually lower-order harmonics because of the predominant effect, it is easy to design a filter for the rest of the harmonics and it can be removed simply. In output voltage, the lower-order harmonics are predominant and dominating to cause the losses which need to be eliminated. In the nine-level inverter, the 5th, 7th and 11th harmonic must be eliminated. It is not required to consider the triple harmonics as they will disappear in a three-phase system. The objective function is written in Equation (1.5) for nine level. In Equation (1.6), *MI* equals modulation index and is provided.

$$\begin{cases} (Cos\,\alpha 1 + \cos\,\alpha 2 + \cos\,\alpha 3 + \cos\,\alpha 4)\,/\,\pi = M \\ (Cos\,5\alpha 1 + \cos\,5\alpha 2 + \cos\,5\alpha 3 + \cos\,5\alpha 4) = 0 \\ (Cos\,7\alpha 1 + \cos\,7\alpha 2 + \cos 7\,\alpha 3 + \cos 7\,\alpha 4) = 0 \\ (Cos\,11\alpha 1 + \cos\,11\alpha 2 + \cos 11\,\alpha 3 + \cos 11\,\alpha 4) = 0 \end{cases} \tag{1.5}$$

$$MI = \frac{V_1}{kV_{dc}} \tag{1.6}$$

1.3   OPTIMIZATION ALGORITHM

1.3.1   Particle Swarm Optimization (PSO)

Particle swarm optimization (PSO) is a random search optimization algorithm in which bird flocking is the motivation of the search mechanism [4]. This algorithm has been applied successfully in many applications. Here, every swarm's fitness value has been assessed by the fitness function which is the same as the THD. The particles shall be flew on the search space—to find the best swarm. In each repetition, each particle has been updated with "pbest" and "gbest" with respect to the velocity. The learning factors c1 and c2 have been selected as 1 and 1.2, respectively. THD shall be assessed in every repetition, for α_k and the "pbest" and "gbest" values shall be updated. The final best THD for switching angles of α_k have been updated using "gbest".

1.3.2   Simulated Annealing (SA)

The SA algorithm is based on the field of chemistry and physics, this process is called annealing [5]. The objective energy function E is the function to be minimized. In the problem, the energy function is THD. The temperature T decreases gradually over the course of the procedure. Δf is the difference between the energies of two states "current_THD" and "new_THD". The probability function depends on T and D, i.e., exp $(-(\Delta f/T))$. THD is the objective of this work.

1.3.3 BAT ALGORITHM (BA)

The echolocation behavior of bats is the inspiration for the development of this algorithm. Generally, the pulse generated by bats lasts for 8–10ms and the range of frequency might be between 25khz and 150khz. The wavelength is 2–14mm [6]. In BA, the following assumptions are made for characterizing the microbat. Pulses of echo and rate of pulse emission out between 0 and 1 confer the location of their food. BA is to be optimized the switching angle of MLI. Initially, the voltage and switching angles are initialized and randomly generated for evaluating the objective function. The objective function of the work is THD minimization. The evaluation of the corresponding switching angles of the system is to be made according to the objective function.

1.3.4 CUCKOO SEARCH ALGORITHM (CSA)

This algorithm imitates the breeding behavior of the cuckoo. The cuckoo bird lays its eggs in the nest of a host bird during its breeding [7–8]. If the host bird identifies the cuckoo egg, then it may abandon the egg. The host bird may abandon its nest and leave to start a new one. When the host bird gets confused due to the similarity of their egg and the cuckoo egg, productivity can be increased. This allows the cuckoo eggs to mature as the number of chicks in the nest is less. Three basic rules are taken for simplification. One egg is considered for each nest. The best nest will be saved for the next generation. The host number of birds is fixed. With this, the survival of the egg can be found by using the probability of Pa, between 0 and 1, which means the possibility of the host bird recognizing the cuckoo egg. In this work, fitness function $f(\alpha)$. The best egg can be found at the location for different switching angles $(\alpha 1, \alpha 2, \alpha 3, \dots, \alpha k)$. In cuckoo search, the controlling parameters are population size and probability (Pa). CSA has been applied to different nonlinear problems. The results obtained are best with the existing algorithm [9–11].

1.4 COMPARISON OF OPTIMIZATION ALGORITHMS

Each algorithm is unique in its searching mechanism and it was developed based on the lifestyle of the kind of species or kind of physical process. Each algorithm is best for the optimization of any nonlinear problem and the comparison is made only related to SHE techniques for multilevel inverters. In this comparison, three kinds of parameters have been considered. They are get struck at local minima, convergence and simulation time. The advantage of PSO is that it finds the best solution in minimum time, but it often gets stuck at local minima. SA does not get stuck at local minima and it is a great advantage of SA. But the disadvantages are that the simulation time taken to converge to global minima is longer than others. Also, it converges at constraints where the constraints are not fully satisfied. Similarly, the BA also sometimes gets stuck at local minima but offers better solutions. CSA is better in converging and does not get stuck at local minima. But the simulation time is a little higher than PSO but less than SA and BA.CSA behave better than others.

1.5   SIMULATION AND EXPERIMENTAL RESULTS

A simulation tool is used in MATLAB which is developed for nine levels. In this, equal voltage sources are selected. The optimization codes as per the flow chart are developed using the MATLAB program. The fitness function $f(\alpha)$ is calculated for different MIs between 0.4 and 1.2. The objective of this work is to reduce fifth, seventh and eleventh harmonics for nine levels. Since, in a three-phase system, triplen harmonics vanish automatically and hence it is considered in this work.

Figure.1.3 shows the output voltage waveform for nine levels. In Figures 1.4–1.7, FFT analysis is shown which is measured practically through a power quality analyzer. From the Figure 1.4, it can be noted that the THD is reduced while using CSA in comparison with others. Table 1.1 shows a comparison of results using the SA, PSO and BA algorithms. From the table, we can see that the line-to-line THD is less for PSO. All the algorithms have satisfied the constraint that all the angles are chosen from 0 to $\pi/2$. Also the angles are not overlaps of each other, this shows that the angles satisfy the constraints, i.e., $\alpha 1 < \alpha 2 < \alpha 3 < \alpha 4$ for nine levels. Figure 1.8

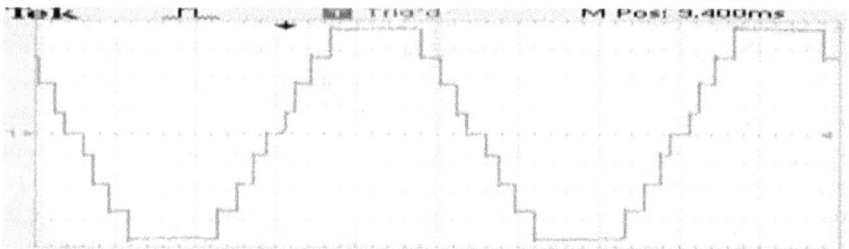

FIGURE 1.3   Nine-level output waveform

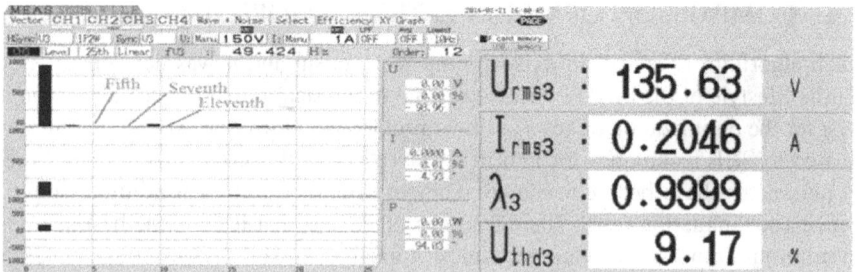

FIGURE 1.4   THD analysis – CSA

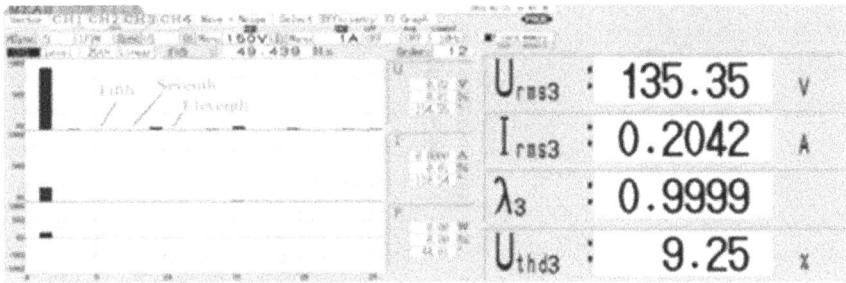

FIGURE 1.5   THD analysis – PSO

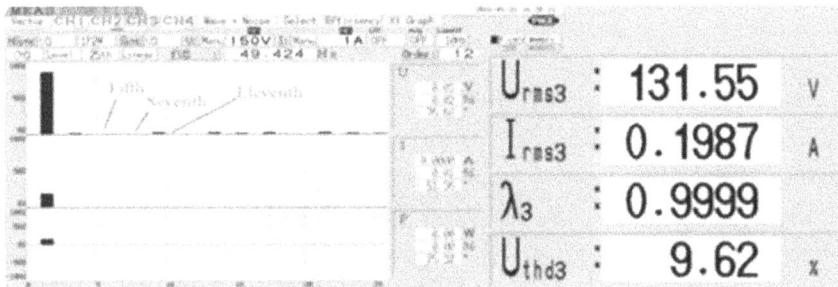

FIGURE 1.6   THD analysis – BA

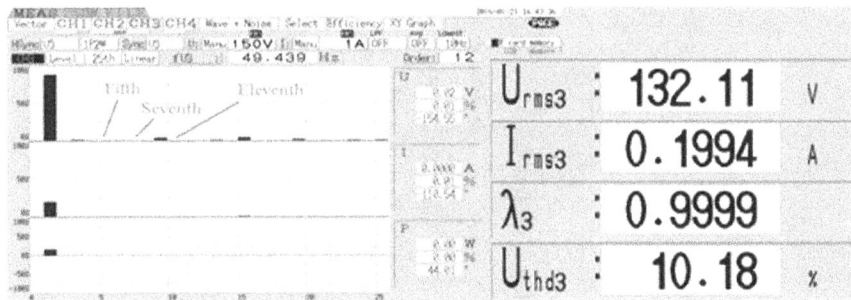

FIGURE 1.7   THD analysis – SA

shows the comparison of individual harmonic order for PSO and CSA which is best out of these four. The harmonics such as fifth, seventh and eleventh are less in CSA. Also up to the thirteenth order of harmonic amplitude, CSA is better than PSO and SA. But in terms of individual harmonics BA provides a much better solution than CSA. In Figure 1.9, the comparative chart between the modulation index and phase THD is shown for all the algorithms used. Similarly, the THD is reduced in CSA over other optimization techniques. Figure 1.10 shows the THD percentage line for

TABLE 1.1
Comparison of algorithms for nine levels

Techniques used	MI	Switching angles by degree				Measured practical THD in % (phase voltage)	Measured THD in % (line voltage)
		α_1	α_2	α_3	α_4		
CSA	0.99	9.132	18.9	34.9	57.1	9.1	5.4
PSO	1.0	9.625	19.308	36.554	59.301	9.23	5.5
BA	0.94	10.026	22.121	40.210	61.764	9.62	6.64
SA	0.96	10.758	24.231	40.483	60.744	10.12	6.09

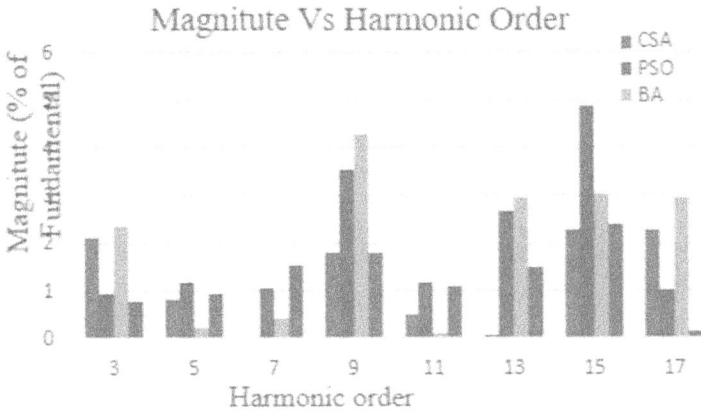

FIGURE 1.8   Comparison of various algorithms

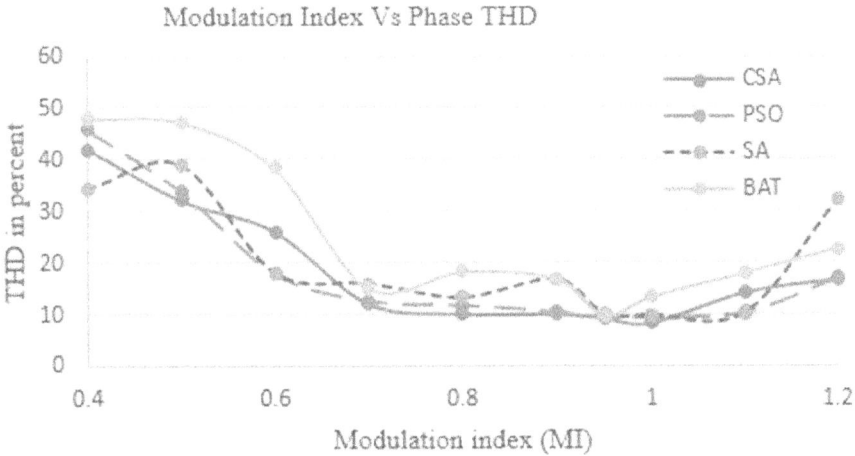

FIGURE 1.9   Modulation index versus phase THD

various modulation index values. From the chart, we can note that the THD is lesser in CSA than in others.

For an experimental setup, MOSFET (20N60C3), power diode (MUR860), MOSFET drivers (TLP250), and isolated DC-DC converters (MINMAX MCW03) have been used to develop a nine-level inverter as shown Figure 1.11. The gate pulse is generated by the controller (ATmega328P).

1.6 CONCLUSION

In this chapter, the classical algorithms such as CSA, PSO, BA and SA have been discussed for optimizing the switching angles of nine-level multilevel inverter. The

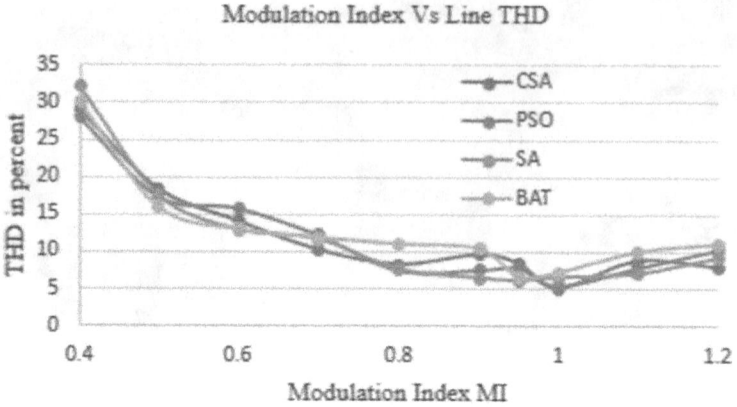

FIGURE 1.10   Modulation index versus line THD

CSA is the best out of all. It provides less THD and the individual lower-order harmonics are also less. The controlling parameters are only two in the case of CSA. They are probability and population size. PSO is often stuck with local minima whereas SA, BA and CSA are not affected by this problem. In the case of SA, it does not get stuck in local minima but the constraints are not satisfied during convergence. The bat algorithm provides better results than SA but the THD is high in comparison with PSO and CSA whereas the lower-order harmonics are less than all others. Hence the overall performance of CSA and BA is much better than others.

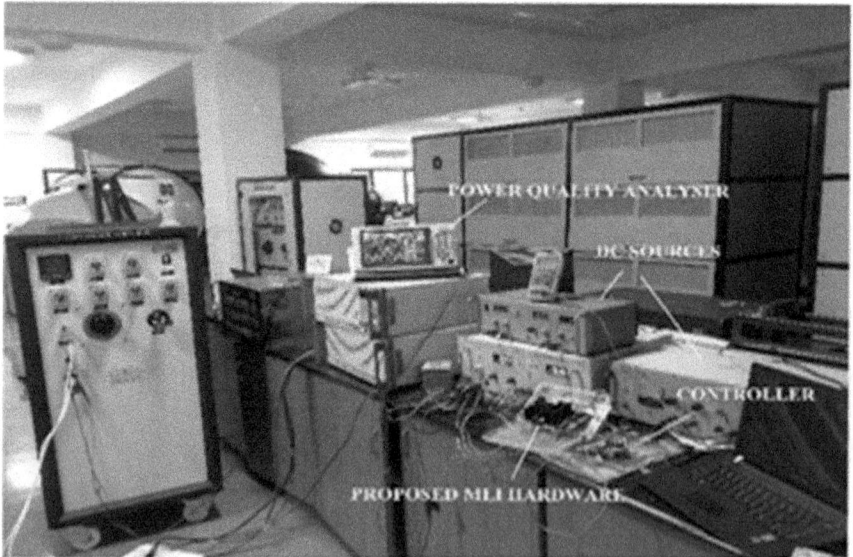

FIGURE 1.11   Experimental setup for nine levels

However, the performance of CSA is much better than that of BA which can be applied to any optimization problem.

REFERENCES

1. J. Roidriguez, S. Bernet, P. K. Steimer, and I. E. Lizama. 2010. A Survey on Neutral-Point-Clamped Inverters. *IEEE Transactions on Industrial Electronics*, 57(7):2219–2230.
2. F. S. Kang, and S. E. Cho. 2005. Multilevel PWM Inverters Suitable for the Use of Stand-Alone Photovoltaic Power System. *IEEE Power Electronics*, 20:906–915.
3. A. Routray, R. K. Singh, and R. Mahanty. 2020. Harmonic Reduction in Hybrid Cascaded Multilevel Inverter Using Modified Grey Wolf Optimization. *IEEE Transactions on Industry Applications*, 56(2):1827–1838.
4. N. V. Kumar, V. K. Chinnaiyan, M. Pradish, and M. S. Divekar. 2015. Enhanced Power Quality of MLI Using PSO Based Selective Harmonic Elimination. International Conference on Green Computing and Internet of Things (ICGCIoT 2015), New Delhi.
5. N. V. Kumar, V. K. Chinnaiyan, M. Pradish, and M. S. Divekar. 2016. An Analysis of Cuckoo Search Algorithm Based Selective Harmonic Elimination for Multilevel Inverter. 2nd International Conference on Intelligent Computing and Application.
6. N. V. Kumar, V. K. Chinnaiyan, M. Pradish, and M. S. Divekar. 2016. Selective Harmonic Elimination: An Comparative Analysis for Seven Level Inverter. IEEE tech-sym 2016.
7. A. Ouaaraba, and X.-S. Yangb. 2014. Discrete Cuckoo Search Algorithm for the Traveling Salesman Problem. *Neural Computing and Applications*, 24(7):1659–1669.
8. H. Zhang, J. Zhou, N. Fang, R. Zhang, and Y. Zhang. 2013. Daily Hydrothermal Scheduling with Economic Emission Using Simulated Annealing Technique Based Multi-objective Cultural Differential Evolution Approach. *Energy*, 50:24–37.
9. J. Islam, S. T. Meraj, A. Masaoud, M. A. Mahmud, A. Nazir, M. A. Kabir, M. M. Hossain, and F. Mumtaz. 2021. Opposition-Based Quantum Bat Algorithm to Eliminate Lower-Order Harmonics of Multilevel Inverters. *IEEE Access*, 9:103610–103626.
10. E. Barbie, R. Rabinovici, and A. Kuperman. 2020. Analytical Formulation and Minimization of Voltage THD in Staircase Modulated Multilevel Inverters With Variable DC Ratios. *IEEE Access*, 8:208861–208878.
11. M. Jafari, and M. R. Islam. 2021. Optimal Design of a Multiwinding High-Frequency Transformer Using Reluctance Network Modeling and Particle Swarm Optimization Techniques for the Application of PV-Linked Grid-Connected Modular Multilevel Inverters. *IEEE Journal of Emerging and Selected Topics in Power Electronics*, 9(4):5083–5096.

2 Design and Development of an Automated Product Retrieval System for a Warehouse

Sadhasivam Udhaykumar, Santhakumar Mohan,
Shanmugham Ashwin Kumar,
Umashankar Jugal Balaji, and Prakash Vishnuvarthan

CONTENTS

2.1 INTRODUCTION

In a supermarket, the vast varieties of products are arranged in an extensive layout of racks. The area is split into sections which contain alike products, such as detergents, cereals, stationery, etc. But even with sufficient directions, people tend to roam about the store searching for the product they need. Automation – a means to increase productivity in a manufacturing plant – can be applied to this problem as well. The time spent by the customer in a supermarket is mainly searching for products and then picking them up. If these actions performed by the customer are automated, then a significant amount of time can be saved. There must also be a means by which the automated system would know which products the customer needs.

An automated guided vehicle (AGV) is a robot which travels along a predefined path performing intended functions along the way. They are normally used in manufacturing plants for logistics operations. Typical uses include retrieval of materials from the store room, transport of materials to the work stations, storing materials in the warehouse, etc. AGVs are so named because they function without the need for instructions from humans. They help to reduce labor costs, increase the efficiency of the system, improve productivity, etc. The first AGVs were introduced during the 1950s. They were guided by a wire which was laid out on the floor. Now we have advanced laser-guided vehicles. A simple AGV consists of a wheeled robot to which a cart or a bed is attached. The components to be transported are placed on the bed and the robot tows the bed to the intended location.

A system in which an AGV picks up the products ordered by the customer (by using a mobile application) from the warehouse and delivers them to the customer waiting at the delivery section of the warehouse. The design of the AGV, the Android application, and the layout traced by the AGV were performed simultaneously. A miniature model was fabricated to simulate the working of the proposed model.

2.2   LITERATURE SURVEY

Xing Wu et al. developed a wheeled robot which possesses omni-directional mobility through a cross-coupling control algorithm to easily maneuver through constricted passages using automated guided driving modules which has a suspension-like mechanism. The robot has a vision system to account for deviation in its omni-directional movement whose parameters are calibrated through a plane target with a grid pattern [1].

Suyang Yu et al. developed an omni-directional AGV with MY3 wheels which provides three complete degrees of freedom in motion planes, a good load-carrying capability, and resistance to slipping. The guiding method of an AGV consists of optical color sensors and onboard cameras which facilitate translational and rotational motion in a specified path [2].

Nicholas Charabaruk et al. developed and tested an autonomous omni-directional robot used for handling radioactive waste with minimal human exposure. The robot was designed with mecanum wheels for providing omni-directional movement, programmed with the open source Robot Operating System (ROS) and linear and rotational odometry tests were performed to test the robot's characteristics [3].

Mostafa Sharifi et al. presented the mechatronic system design and development of a four-wheel drive/steer (4WD4S) mobile robot as a nonholonomic omni-directional robot, MARIO – Mobile Autonomous Robot for Intelligent Operations – and implementation of innovative integrated applications CAD/CAM/CAE and RP in the rapid development of the robot chassis and other mechanical parts by using different software tools [4].

Shuai Guo et al. developed an omni-directional industrial robot drilling system for aerospace manufacturers and applied a vision system to improve the precision of mobile drilling. In this system, a laser measurement system and displacement measurement system are used to do the autonomous navigation and anti-collision job, and finally, experiments were carried out to compare the influence of different calibration targets on the robot system [5].

Michiko Watanbe et al. created a sparse distributed memory neural network (SDMNN) coupled with Q-learning to avoid the problems incurred in the navigation and operation of AGVs by recognizing the acquired scenes and by the method of mutual understanding through Q-learning while multiple AGVs are operating simultaneously [6].

K.R.S Kodagoda et al. proposed a technique for lateral and longitudinal control of vehicles. It focuses on the development and implementation of intelligent and stable fuzzy Proportional Derivative Proportional Integral (PDPI) for speed control and steering of the AGV [7].

Riccardo et al. discussed a decision support system which implements a Database Management System (DBMS) and Graphical User Interface (GUI) for the design and management of warehouse systems through a top-down methodology wherein the main decision steps are layout, allocation and assignment based on which alternative warehouse configurations are chosen through a what-if multi-scenario simulation [8].

Shengjun Shao et al. implemented a semaphore-based traffic control model for synchronizing concurrent AGVs in operation to resolve cross conflicts and head-on conflicts through SMSL (Synchronize Multiple AGVs in a Segmented Lane) protocol and to prevent deadlocks through SMCL (Synchronize Multiple AGVs in a Continuous Lane) protocol [9].

Luo et al. proposed a design of a Programmable Logic Controller (PLC) based on Petri Net (PN) to avoid collision between vehicles in the system. The control-structure design was made for line, merge, divide and intersection alignments and the closed loop PN was translated to a Ladder Logic Diagram (LLD) using Boolean expression for the PLC [10].

Abdullah-Al-Mashud et al. proposed a system with a remote vehicle which is capable of traveling in a preprogrammed path without any human intervention. A personal computer is interfaced with the remote vehicle through a parallel printer port for which input is fed in a serial binary form [11].

Luiz Felipe Verpa et al. simulated a production line consisting of AGVs with real-time data using Promodel 7.0 to optimize the material flow for increasing productivity and inferred that the system has been utilized to maximum capacity with minimum idle time [12].

Cotet et al. used CAD to develop a virtual model of an Automated Storage and Retrieval System (AS/RS) and Rail Guided Vehicles (RGVs) beginning from 3D modeling and ending with platform testing and integration with its manufacturing cell followed by performance diagnosis through witness simulation software [13].

Suman Kumar Das et al. discussed the design and different methodologies of AGV systems and provides an overview of AGVs' technological developments and describes the formulation to control the traffic inside industrial workspace, floor control and traffic management systems [14].

Farah Hanani M.K. et al. studied AS/RS features and operating procedures and assessed the related hardware, software and communication modules and explored the choices of several hardware and software modules available in the current market and discussed AS/RS design considerations and the limitations faced by the designers during the process of project development and implementation [15].

Robert Harrison et al. reviewed the engineering approaches that have been proposed or adopted to date, including Industry 4.0 and provided examples of engineering methods and tools that are currently available. They concentrated on the Cyber-Physical Systems (CPSs) engineering toolset being developed by the Automation Systems Group (ASG) and explained via an industrial case study how such a component-based engineering toolset can support an integrated approach to virtual and physical engineering of automation systems [17].

Shaoping Lu et al. discussed Radio Frequency Identification (RFID) enabled positioning system in AGV for smart factories and examined the key factors on AGVs accuracy such as a magnetic field in circular antenna, a circular magnetic field and circular contour stability quantitatively, and performed a simulation study and discussed the results on the basis of the diameter of the antennas used in parking and driving zones [18].

Vishakha Borkar et al. developed a pick–and-place robotic arm which is interfaced with a GUI application. The pick-and-place unit is programmed with the help of an AVR microcontroller and the wireless parts were validated by Cyclic Redundancy Check (CRC) error detection [19].

Omijeh B.O. et al. have validated the structural stability of a remote-controlled robotic vehicle's manipulator through static and multi-body dynamic analysis. The electrical circuits in the robotic unit are validated through Channels Circulation Analysis (CCA) for checking the correctness of pulse signals provided to the servo motor [20].

Francis Bourbonnais et al. created an optimization algorithm by implementing a cubic spline stochastic approach which reduces the trajectory time. By bypassing all parallel singularities, the suitable combination for a working region was identified to increase the size of the workspace in a five-bar parallel pick-and-place robot [21].

It is evident from the literature survey that omni-directional wheels are a topic of interest in AGV design over recent years. Limited work was found specifically for AGV control. However extensive works have been found in the design of AGV systems as a whole. The use of CAD/CAM software for modeling and simulation is extensive. RFID technology is still being worked on. Newer concepts like cyber-physical systems are also being explored. As far as robotic arms are concerned, most works are related to optimizing the trajectory of motion. Any innovation in future works may be oriented toward omni-directional motion of the AGV, implementation of Internet of Things (IoT) for AGV, arm control and monitoring, cyber-physical systems, artificial intelligence for AGV and arm functioning, etc. Though AGV and robotic arms have been extensively used in industrial applications to optimize workspace and flow of materials, their application to commercial facilities such as grocery stores is scarce.

2.3   CONCEPTUAL DESIGN

The idea behind the conceptual design was to mount a robotic arm atop an AGV, which adds the pick-and-place function of the robotic arm and the navigation function of the

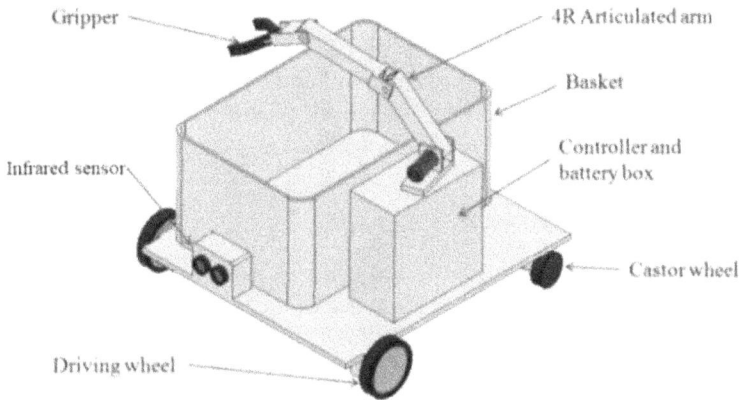

FIGURE 2.1   Conceptual design of the AGV

AGV, thereby forming a device which will help meet the objective. A 4R articulated arm was chosen so that the end effector orientation can be maintained at 0°. This is necessary when handling liquid containers as they may start leaking if tilted too much. A basket is provided in which the picked products can be kept. In the front of the vehicle, sensors will be mounted to aid the robot in navigation. According to the conceptual design, the vehicle will navigate itself by dead reckoning. However, owing to inaccurate movements observed during the testing of the fabricated model, the vehicle was made into a line follower. The vehicle has two driven wheels in the front and two castor wheels at the rear. The CAD model of the conceptual design is shown in Figure 2.1.

The conceptual design for the racks is shown in Figure 2.2. The dimensions of the rack are 2m × 1m × 2.5m and the lowermost row is 0.5m above the ground.

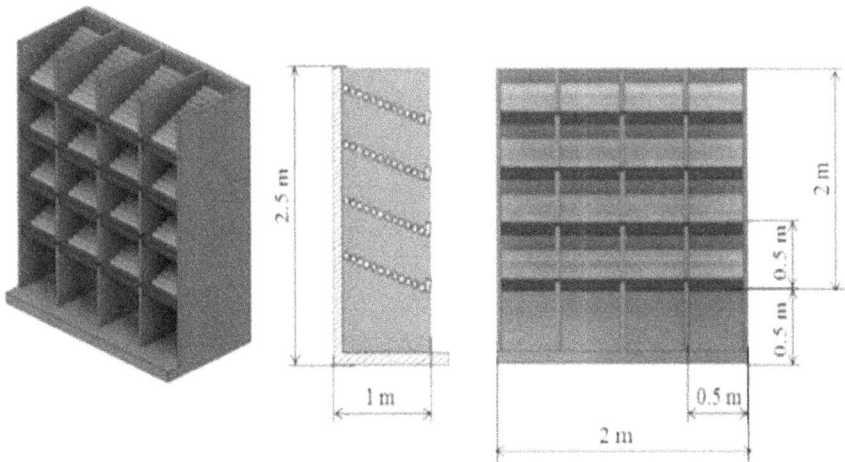

FIGURE 2.2   Conceptual design of the rack

The racks were designed in such a way that when one unit of product is taken, the next unit occupies the former's position. This is accomplished by inclining the base of the cells in the rack as shown in Figure 2.2. By doing so, the position at which the end effector must be brought to pick a particular product always remains the same.

2.4   DETAILED DESIGN OF THE ARM

The arm is the most important as well as complex part of the vehicle; therefore, a detailed design was made for it. In order to decide the length of the links for the arm, the range of the arm must be known. To find the range of the arm, the positions at which the products will be kept are required. The coordinates at which the end effector must be positioned for picking the products from the rack are shown in Figure 2.3. The cells in the rack occupy a square of dimensions 2m × 2m as shown in Figure 2.2. The XZ plane was taken parallel to the front side of the rack at a distance of 200mm from the rack. The origin was fixed coincident with the center of the square occupied by the cells as shown in Figure 2.3.

A schematic representation of the 4R articulated arm is shown in Figure 2.4. These notations will be used for all the calculations.

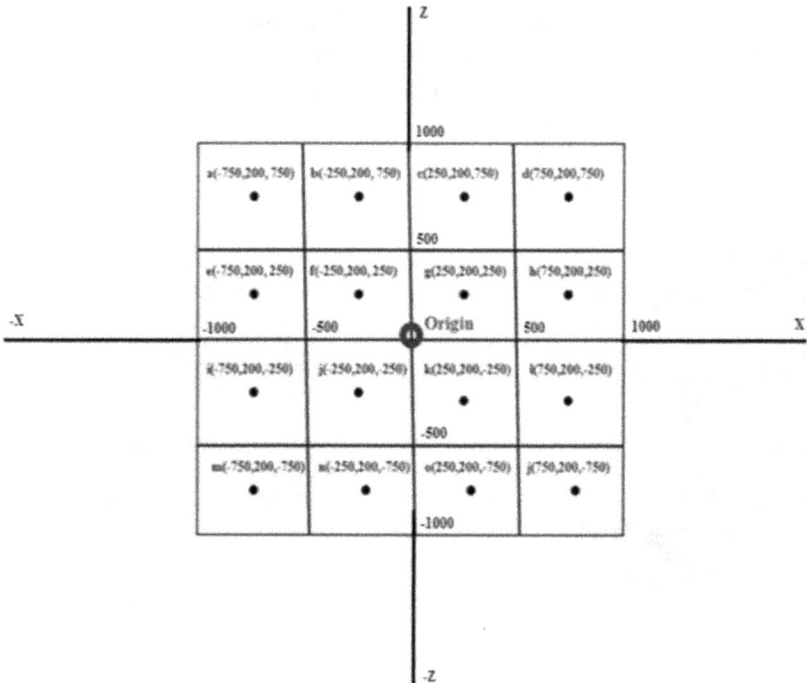

FIGURE 2.3   Coordinates to pick products

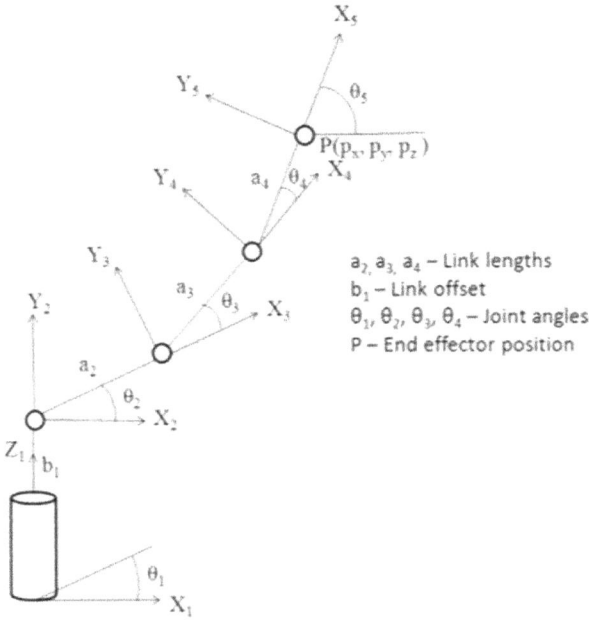

FIGURE 2.4 Schematic representation of the arm

From Figure 2.4, the extreme position that the end effector must reach is (750, 200, 750). The distance between the origin and this extreme position must be the minimum length of the arm. The distance can be calculated as:

$$= \sqrt{\left(x_2 - x_1\right)^2 + \left(y_2 - y_1\right)^2 + \left(z_2 - z_1\right)^2} \qquad (2.1)$$

Where, $\left(x_1, y_1, z_1\right) = \left(0,0,0\right)$

$$\left(x_2, y_2, z_2\right) = \left(750, 200, 750\right) = \sqrt{\left(750\right)^2 + \left(200\right)^2 + \left(750\right)^2} = 1079.3\,\text{mm}$$

Based on the minimum length required the lengths of the links a_2, a_3, and a_4 were determined and are listed in Table 2.1.

TABLE 2.1
Length of the links of the arm

Link	Length (mm)
a_2	500
a_3	500
a_4	200

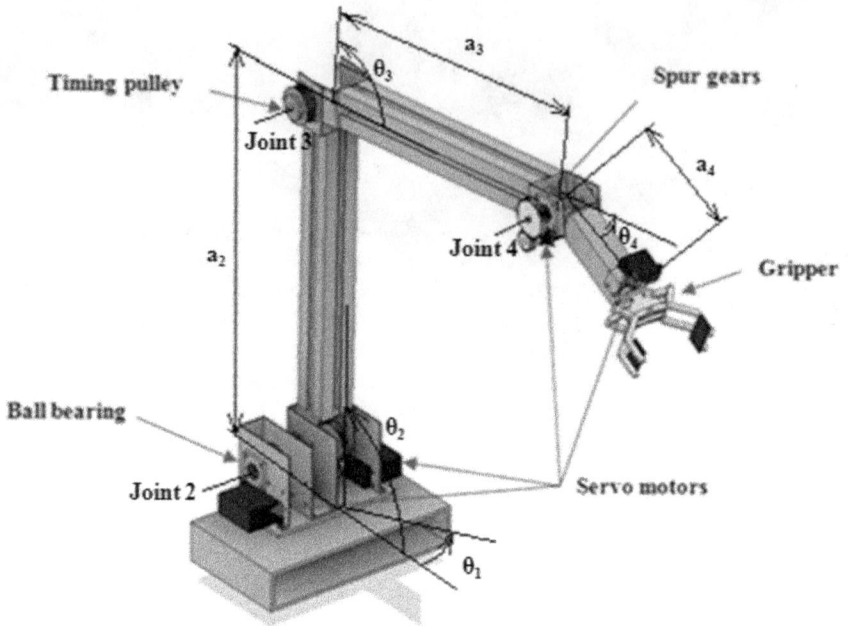

FIGURE 2.5   Detailed design of the arm

The detailed design of the arm was made with the determined link lengths. The CAD model of the detailed design is shown in Figure 2.5.

DC servo motors drive the joints of the arm. Spur gears are used to transmit power from the motors to the joint and to get the required torque and speed. The motor for driving the third joint is connected to the joint through a timing belt so that the motor can be placed close to the base of the arm. This helps to reduce the torque at the second joint. A parallel gripper is used here which facilitates easier holding of boxes regardless of their size. The gripper works on the parallelogram principle.

The body of the vehicle is very simple with no moving parts as there is no suspension, differential, or any other complex mechanisms. Differential steering is achieved by driving the motors at different speeds; the speeds being controlled electronically. Owing to its simplicity of design as well as the full-scale model not being fabricated, the design of the body was not detailed upon.

2.5 INVERSE KINEMATICS

The microcontroller controls the rotation of the servos which in turn controls the position of the end effector of the arm. Therefore, the inputs needed to control the position of the end effector are the joint angles. The process of finding the joint angles for a given position of the end effector is called inverse kinematics. The inverse kinematic analysis for the 4R articulated arm is discussed as follows. (Note: $\sin x$ and $\cos x$ are abbreviated as s_x and c_x respectively for convenience.)

Let the end effector position be (p_x, p_y, p_z). The joint angle θ_1 can be found as:

$$\theta_1 = atan2(p_x, p_y) \tag{2.2}$$

The other solution for θ_1 is:

$$\theta_1 = \pi + atan2(p_x, p_y) \tag{2.3}$$

After θ_1 is determined, we can say that the arm is planar with respect to variables θ_2, θ_3 and θ_4. It is easier to derive the joint angles of a 2R planar arm, therefore the position of the wrist is determined from the end effector position (p_x, p_y, p_z), orientation angle φ and θ_1, which makes it equivalent to a 2R arm. If the wrist position is denoted as (w_x, w_y, w_z) then it can be calculated as:

$$w_x = p_x - a_4 c_\varphi c_1 \tag{2.4}$$

$$w_y = p_y - a_4 c_\varphi c_1 \tag{2.5}$$

$$w_z = p_z - a_4 s_\varphi \tag{2.6}$$

Now, c_3 and s_3 can be determined as:

$$c_3 = \frac{w_x^2 + w_y^2 + w_z^2 - a_2^2 - a_3^2}{2 a_2 a_3} \tag{2.7}$$

$$s_3 = \pm\sqrt{1 - c_3^2} \tag{2.8}$$

The value of θ_3 can be calculated from c_3 and s_3 as:

$$\theta_3 = atan2(s_3, c_3) \tag{2.9}$$

Then, c_2 and s_2 can be determined as:

$$c_2 = \frac{(a_2 + a_3 c_3)\sqrt{w_x^2 + w_y^2} + a_3 s_3 w_z}{w_x^2 + w_y^2 + w_z^2} \tag{2.10}$$

$$s_2 = \frac{(a_3 + a_3 c_3) w_z - a_3 s_3 \sqrt{w_x^2 + w_y^2}}{w_x^2 + w_y^2 + w_z^2} \tag{2.11}$$

The value of θ_2 can be calculated from c_2 and s_2 as:

$$\theta_2 = atan2(s_2, c_2) \tag{2.12}$$

```
Python 3.7.2 Shell                                                          _  □  ☒
File  Edit  Shell  Debug  Options  Window  Help
Python 3.7.2 (tags/v3.7.2:9a3ffc0492, Dec 23 2018, 22:20:52) [MSC v.1916 32 bit
(Intel)] on win32
Type "help", "copyright", "credits" or "license()" for more information.
>>>
 RESTART: C:\Users\lenovo\AppData\Local\Programs\Python\Python37-32\Invere kinem
atic.py
Enter end effector orientation angle in degree0
Enter end effector coordinates
px = 250
py = 200
pz = 750
Solution 1
38.65980825409009 41.277803476279644 79.8128266535508 -121.09063012983044
Solution 2
38.65980825409009 117.55522656082226 -79.8128266535508 -37.74239990727146
>>>
```

FIGURE 2.6 Program output

Finally, θ_4 can be calculated as:

$$\theta_4 = \varphi - \theta_2 - \theta_3 \tag{2.13}$$

Thus, for any given end effector position (p_x, p_y, p_z), the joint angles can be calculated using the above procedure.

Since the equations are time consuming to solve manually, the equations were implemented as a Python program. The program takes the end effector position and the link lengths as its input, giving the joint angles as the output. The output of the program for the input $(p_x, p_y, p_z) = (250, 200, 750)$ is shown in Figure 2.6.

As mentioned earlier in this section, for any given possible position of the end effector, there exist two solutions, i.e., two different sets of joint angles. A schematic representation of the solutions from the program output is shown in Figure 2.7.

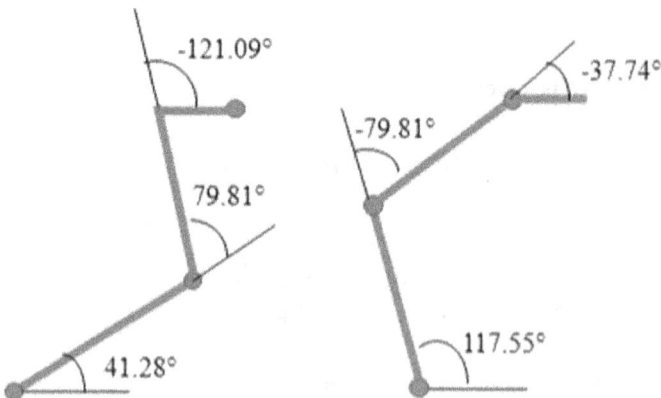

FIGURE 2.7 Schematic representations of the solutions (a) Solution 1 (b) Solution 2

2.6   FABRICATED MODEL

For testing the feasibility of the designs, a scaled-down model was fabricated. The scaled-down model was built to have all the functionality of the full-scale robot while also keeping the cost low. For example, cheap micro servos were used in the fabricated model and the arm size was reduced accordingly to suit the capacity of the servos. The fabricated model is shown in Figure 2.8.

The base, basket and platform for mounting the arm and the links of the arm were all made from aluminum sheets of thickness 2mm. The pieces were cut from the sheet and then stuck to each other using hot-melt glue with further reinforcement with adhesive tape. The dimensions of the frame and the arm are shown in Figure 2.9.

The other components were mounted on the aluminum sheets by using double-sided tape, hot-melt glue, or instant adhesive. The components used are listed in Table 2.2.

The robot navigates by following a black line present on the floor. Three infrared sensors are used for detecting the black line. They make use of the fact that a black surface does not reflect infrared light. The line-following logic is shown in Figure 2.10.

The layout through which the robot will navigate itself is shown in Figure 2.11. The stations, home and delivery section are recognized with the help of a rectangle of black which can absorb light from all three sensors. The robot starts at "Home" and would stop at a station if a product has to be picked up at that station. After picking up all the required products the robot delivers them to the

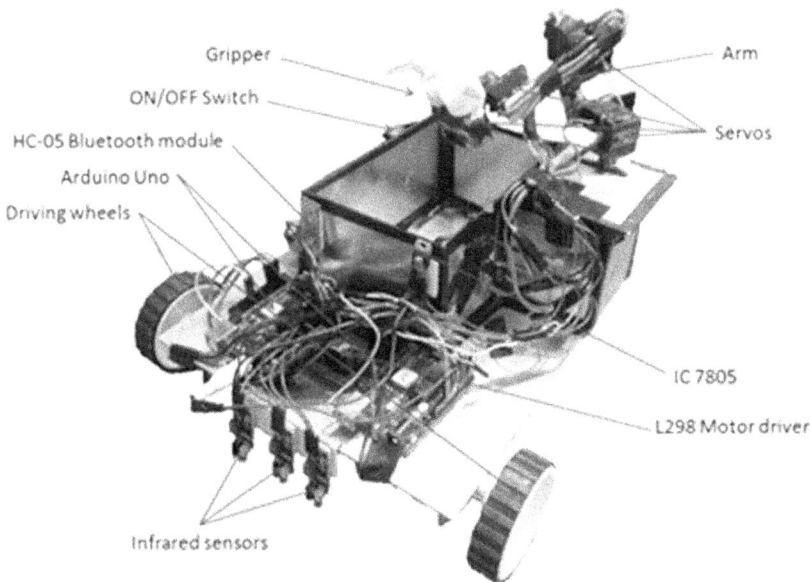

FIGURE 2.8   Fabricated model

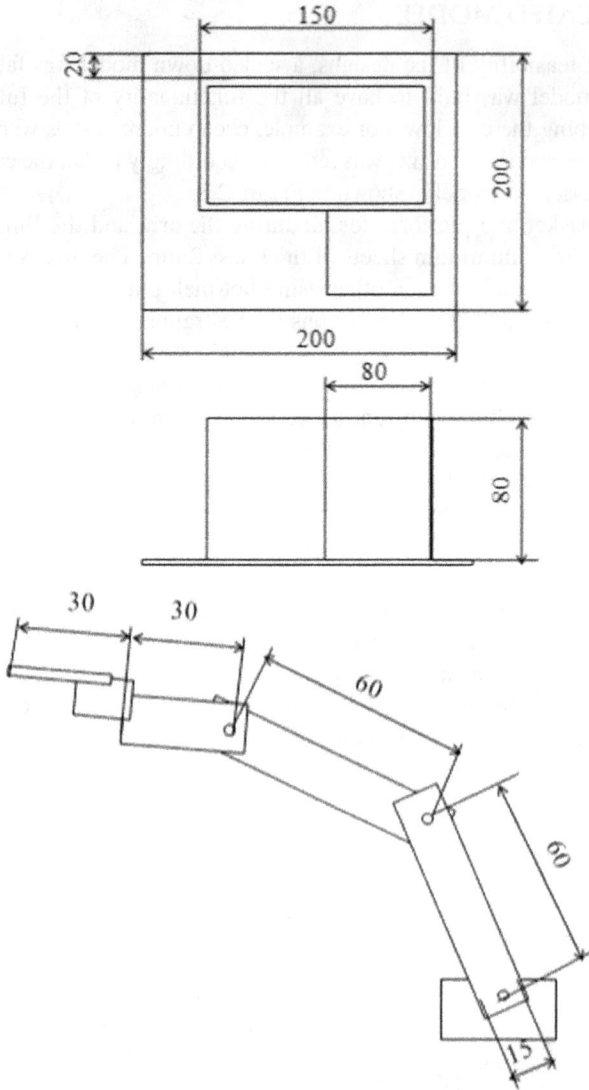

FIGURE 2.9 Dimensions (a) frame (b) arm

delivery section and returns back "Home". All these functions are controlled by one of the Arduino boards.

The movements of the arm are controlled by the other Arduino board. It sends signals to the servos and makes them rotate to the required position. The arm begins to function when the robot reaches a station in which a product has to be picked up. The stations are detected with the same three infrared sensors used by the line follower, as they are connected to this Arduino board as well.

TABLE 2.2

List of components

Components	Quantity
Arduino uno	2
L298 motor driver	1
HC – 05 Bluetooth module	1
IC 7805	1
100μF capacitor	1
2 kΩ resistor	5
1 kΩ resistor	3
Wheel	2
Castor wheel	2
Spur gear	2
Servo motors	5
Geared DC motor	2
1.5V dry cell	8
Infrared sensor	3
Connecting wires	As required

MOVES FORWARD   TAKES LEFT TURN   MOVES FORWARD

FIGURE 2.10 Line-following logic

The products which must be picked up by the robot are selected by using an Android application. It is a simple one-screen interface with a list of available products and the user can select the number of units of a product he/she needs. A screenshot of the application is shown in Figure 2.12. First, the smartphone is connected to the HC-05 Bluetooth module by pressing the "Connect" button and selecting HC-05 from the paired devices. Once connected successfully, the label below "Connect" button displays "Connected" in the color green. Then the quantity of each product is selected from their respective drop-down menus. Finally, the "Order" button is clicked and it displays "Order confirmed".

The data from the Android application is received by the HC-05 module. The module is connected to the Arduino board which controls the arm. However, the data must also be available for the other Arduino board as it must know at which stations to stop. For transferring data from one Arduino board or another, an I2C (Inter-Integrated Circuit) connection was used. The Arduino board connected to the Bluetooth module acts as the master and the other Arduino board acts as the slave.

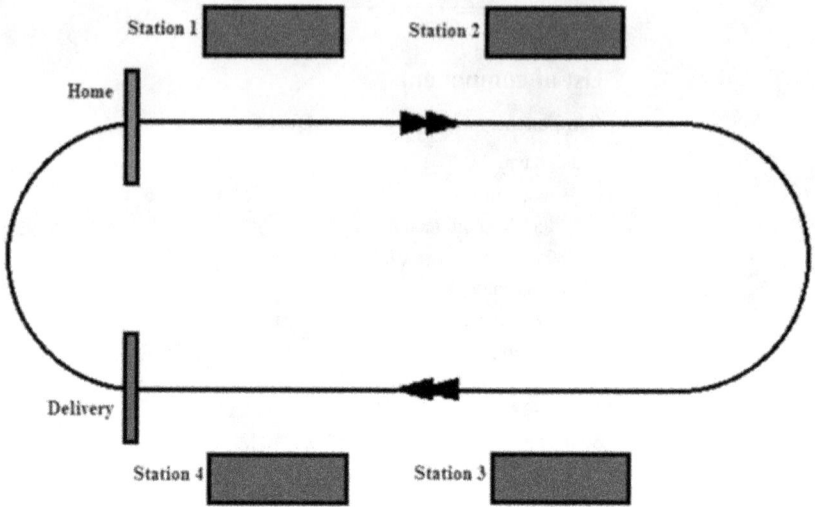

FIGURE 2.11   Robot navigation layout

FIGURE 2.12   Android application

2.7   ALGORITHM FOR PATH FINDING

The layout used in the demonstration of the scaled-down model is highly simple and would be impractical in a real-world scenario. To provide an insight into how the proposed solution could work in a scaled-up layout, an algorithm for the navigation of the AGV has been suggested. Consider a layout in which the racks

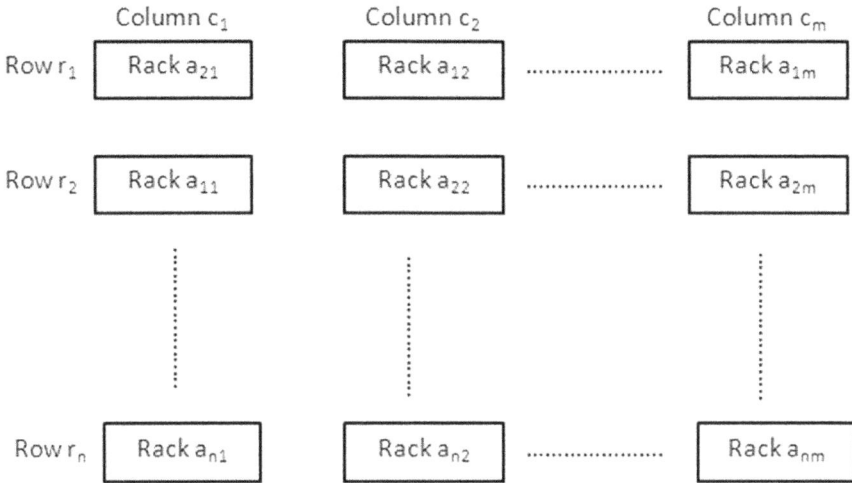

	Column c_1		Column c_2		Column c_m
Row r_1	Rack a_{21}		Rack a_{12}	....	Rack a_{1m}
Row r_2	Rack a_{11}		Rack a_{22}	....	Rack a_{2m}
Row r_n	Rack a_{n1}		Rack a_{n2}	....	Rack a_{nm}

FIGURE 2.13   An array of racks

are arranged in a matrix with "n" number of rows and "m" number of columns as shown in Figure 2.13.

Let r_n and c_m be the nth row and m^{th} column respectively. a_{nm} is a rack in row r_n and c_m. There will be total $n \times m$ racks in the layout. If a matrix is formed comprising the length of the paths connecting every rack to every other rack, it will consist of $(n \times m)^2$ elements. This matrix can be solved using the algorithm given below to find out the shortest path to go to each station and return to the home position. The algorithm that has been used here is an implementation of the "travelling salesman problem" wherein the goal is to find the shortest path to visit a set of cities and return back to the starting point. The algorithm is shown in the flowchart in Figure 2.14.

However, it is not always needed to navigate through all the racks since it is sufficient to go to the racks which contain the products that the customer has ordered. Hence, a matrix which consists of the lengths of the paths connecting those racks which contain the products ordered by the customer is sufficient.

For example, consider a simple layout wherein there are three rows of racks with three racks in each row. This layout is shown in Figure 2.15.

The distance between each rack is 10 units, both row-wise and column-wise. The distance to be traveled from each rack to every other rack has been given in Figure 2.16.

It can be seen that the table is in the form of a matrix. Say, the customer orders products which are in rack numbers 1, 2, 6, 8, and 9. A 5×5 matrix can be formed from the larger matrix by taking only the cells corresponding to racks 1, 2, 6, 8, and 9. This matrix was solved using the algorithm mentioned earlier to find out the shortest path. The shortest path was returned as [1, 6, 2, 8, 9] whose path length is 80 units.

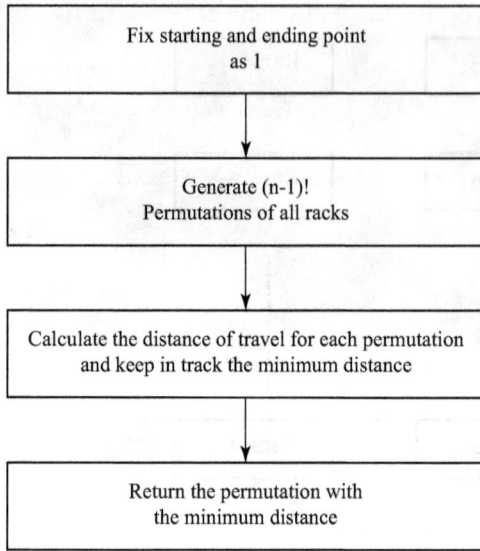

FIGURE 2.14   Flowchart for finding the shortest path

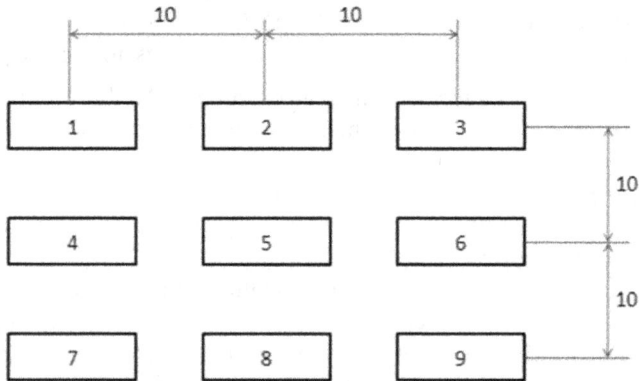

FIGURE 2.15   Layout example

2.8 VALIDATION TECHNIQUES

Two methods which can improve the accuracy of the system have been briefly discussed as follows. To ensure that the correct product is being picked, a camera can be fixed to the end effector. The image of the product as captured by the camera can be compared with the image of the same product stored in the database by using image processing and analysis techniques. Image recognition technology, which uses deep learning, could also be used alternatively to further improve the validation accuracy. The weight of the picked product can also be checked to detect any weight loss due

Rack no.	1	2	3	4	5	6	7	8	9
1	-	10	20	10	20	30	20	30	40
2	10	-	10	20	10	20	30	20	30
3	20	10	-	30	20	10	40	30	20
4	10	20	30	-	10	20	10	20	30
5	20	10	20	10	-	10	20	10	20
6	30	20	10	20	10	-	30	20	10
7	20	30	40	10	20	30	-	10	20
8	30	20	30	20	10	20	10	-	10
9	40	30	20	30	20	10	20	10	-

FIGURE 2.16   Distance between different racks in the example layout

to defective packaging. This can be accomplished by using a torque sensor at joint two. The torque at each joint for a given force at the hand frame of the robotic arm can be determined using Equation (2.14).

$$\left[T\right]=\left[{}^{H}J\right]^{T}\left[{}^{H}F\right] \tag{2.14}$$

Where,

$\left[T\right]$ consists of the torques (for revolute joint) and forces (for prismatic joint) at each joint

$\left[{}^{H}J\right]$ is the Jacobian matrix and

$\left[{}^{H}F\right]$ consists of forces and moments acting in the x, y, and z axes of the hand frame

2.9 CONCLUSION

An attempt was made to reduce the time spent by the customer procuring products from a supermarket by designing an AGV coupled with a robotic arm. A detailed design of the robotic arm was made for picking up products from proprietary racks with inclined cells. The length of the links was determined based on the extreme position the end effector should reach to pick a product. The link lengths are listed in Table 2.3.

Inverse kinematic analysis was performed for the robotic arm to determine the joint angles for the required end effector positions. The inverse kinematic equations were implemented in a Python program. The set of joint angles found for one of the end effector positions (250, 200, 750) is tabulated in Table 2.4.

TABLE 2.3

Length of the links of the arm

Link	Length (in mm)
a_2	500
a_3	500
a_4	200

TABLE 2.4

Joint angles

Joint angle	Solution 1 (in degrees)	Solution 2 (in degrees)
θ_1	38.65	38.65
θ_2	41.28	117.55
θ_3	79.81	−79.81
θ_4	−121.09	−37.74

REFERENCES

1. Wu, Xing, Linjun Zhu, Chenchen Shi, and Peihuang Lou. "Development of an omnidirectional mobile robot with two automated guided driving modules." In *Robotics and Biomimetics (ROBIO), 2014 IEEE International Conference on*, 2014, pp. 1775–1780. IEEE.
2. Yu, Suyang, Ye Changlong, Hongjun Liu, and Jun Chen. "Development of an omnidirectional Automated Guided Vehicle with MY3 wheels." *Perspectives in Science* 7 (2016): 364–368.
3. Charabaruk, Nicholas, and Scott Nokleby. "Design and development of an autonomous omnidirectional hazardous materials handling robot." *Transactions of the Canadian Society for Mechanical Engineering* 40(2) (2016): 169–190.
4. Sharifi, Mostafa, Matthew Samuel Young, X. Chen, and Don Clucas. "Mechatronic design and development of an omnidirectional mobile robot for automation of primary production." In *International Conference on Innovative Design and Manufacturing (ICIDM), Auckland, New Zealand*, 2016, pp. 24–26.
5. Guo, Shuai, Qizhuo Diao, and Fengfeng Xi. "Vision based navigation for omni-directional mobile industrial robot." *Procedia Computer Science* 105 (2017): 20–26.
6. Watanabe, Michiko, Masashi Furukawa, and Yukinori Kakazu. "Intelligent AGV driving toward an autonomous decentralized manufacturing system." *Robotics and Computer-Integrated Manufacturing* 17(1–2) (2001): 57–64.
7. Wijesoma, W. S., K. R. S. Kodagoda, and Eam Khwang Teoh. "Stable fuzzy state space controller for lateral control of an AGV." *Journal of VLSI Signal Processing Systems for Signal, Image, and Video Technology* 32(1–2) (2002): 189–201.
8. Accorsi, Riccardo, Riccardo Manzini, and Fausto Maranesi. "A decision-support system for the design and management of warehousing systems." *Computers in Industry* 65(1) (2014): 175–186.

 9. Shao, Shengjun, Zeyang Xia, Guodong Chen, Jun Zhang, Ying Hu, and Jianwei Zhang. "A new scheme of multiple automated guided vehicle system for collision and dead-lock free." In *2014 4th IEEE International Conference on Information Science and Technology*, 2014, pp. 606–610. IEEE.
10. Luo, Ji Liang, Hui Juan Ni, and Meng Chu Zhou. "Control program design for automated guided vehicle systems via Petri nets." *IEEE Transactions on Systems, Man, and Cybernetics: Systems* 45(1) (2015): 44–55.
11. Mashud, M. A. A., M. R. Hossain, Mustari Zaman, and M. A. Razzaque. "PC guided automatic vehicle system." arXiv Preprint ArXiv:1501.01109, 2015.
12. Leite, Luiz Felipe Verpa, Robson Marinho A. Esposito, Ana Paula Vieira, and Fabio Lima. "Simulation of a production line with automated guided vehicle: A case study." *Independent Journal of Management & Production* 6(2) (2015): 269–285.
13. Cotet, C. E., C. L. Popa, G. Enciu, A. Popescu, and T. Dobrescu. "Using CAD and flow simulation for educational platform design and optimization." *International Journal of Simulation Modelling (IJSIMM)* 15(1) (2016): 5–15.
14. Das, Suman Kumar, and M. K. Pasan. "Design and methodology of automated guided vehicle-A review." *IOSR Journal of Mechanical and Civil Engineering (IOSR-JMCE)* (2016): 29–35.
15. MK, Farah Hanani, M. Y. Zulkhairi, and Mohamad Zaihirain MR. "Development of automated storage and retrieval system (ASRS) for flexible manufacturing system (FMS)." *Journal of Engineering Technology* 4 (2016): 43–50.
16. Harrison, Robert, Daniel Vera, and Bilal Ahmad. "Engineering methods and tools for cyber–physical automation systems." *Proceedings of the IEEE* 104(5) (2016): 973–985.
17. Lu, Shaoping, Chen Xu, and Ray Y. Zhong. "An active RFID tag-enabled locating approach with multipath effect elimination in AGV." *IEEE Transactions on Automation Science and Engineering* 13(3) (2016): 1333–1342.
18. Borkar, V. "Development of pick and place robot for industrial." *International Research Journal of Engineering and Technology(IRJET)* 4(9) (2017): 347–356.
19. Omijeh, B. O., R. Uhunmwangho, and M. Ehikhamenle. "Design analysis of a remote controlled pick and place robotic vehicle." *International Journal of Engineering Research and Development* 10(5) (2014): 57–68.
20. Saha, Subir Kumar. *Introduction to Robotics*. New Delhi: McGraw Hill Education (India) Private Limited, 2014.
21. Niku, Saeed B. *An Introduction to Robotics Analysis, Systems, Applications*. Upper Saddle River, NJ: Prentice Hall, 2001.
22. Singh, N. "Traveling salesman problem (TSP) implementation - GeeksforGeeks [online]." *GeeksforGeeks*, n.d. Available at: https://www.geeksforgeeks.org/traveling-salesman-problem-tsp-implementation/.

3 Implementation of Water and Air Quality Monitoring System for Smart Agriculture using LoRa

Adarsh V Srinivasan, N. Saritakumar,
V. Amarnath, and Amrithaa Sivakumar

CONTENTS

3.1 INTRODUCTION

The existing IoT methods make use of Wi-Fi, Bluetooth, and NFC to communicate between various devices. All these methods have either high power consumption or very short range. It is known that Wi-Fi and Bluetooth are limited in range (less than 15m) and can consume a lot of battery. For an end device to be made, these limitations will pose degradation in terms of the performance of the device. For instance, if sensors are to be across a field, the following are required:

1. Long-range communication
2. Larger power source

DOI: 10.1201/9781003307716-3

Therefore, the existing methods cannot be adopted. Recent technologies concentrate more on increasing the data rate, but there hasn't been the same progress on the distance that the data can travel. For a medium-sized house, Wi-Fi performs efficiently but, when the distance increases separate routers have to be provided which will be inefficient in terms of power and the number of devices used. But this trade-off in power is acceptable for large data [1]. In case of small data transfer over large distances, Long Range Wide Area Network (LoRaWAN) systems will prove to be a reliable solution [2]. LoRa is a technology developed by a company called Semtech which is a new wireless protocol designed mainly for providing long-range, low-power communications [3]. LoRa stands for Long Range Radio and is predominantly used for IoT devices and IoT networks. A number of applications that are running on the same network can be connected using public networks through this technology.

A single LoRa gateway can handle hundreds of nodes at any given instant. The signals may span over a significant distance. Thus, this technology requires less infrastructure as it makes the implementation faster and the construction of a network cheaper [4].

LoRa is a method which uses Chirp Spread Spectrum (CSS) technology for transmitting radio signals. Because of the use of CSS, the data are encoded and hence security is improved. Essentially, these chips are of the Industrial, Scientific and Medical (ISM) band that can facilitate LoRa for converting radio frequency to bits. This technology can be implemented in a physical layer that may be employed in several types of applications even outside of a wide area network [5]. LoRa technology enables connecting several sensors, gateways, machines, devices, etc., wirelessly to the cloud. LoRa makes use of an adaptive data rate algorithm which helps to maximize network capacity and the nodes' battery life [6]. The LoRa protocol includes different layers including encryption at the network, application, and device level which helps in implementing secure communication.

3.2   OBJECTIVES

To design and implement a solution for smart agriculture, water and air quality monitoring for smart cities using LoRa which has the following features:

- Create a low-power solution
- Maximize the area covered with a minimal number of sensors
- Automation of the process

3.3   BLOCK DIAGRAM

As shown in Figure 3.1, the Arduino will be used to power the LoRa shield and the various sensors connected with the shield. The automation process is also done using the data received at the shield and Arduino. Further, the data from the shield is received by the gateway and then sent to the cloud. The data from the cloud can be viewed on a variety of end devices ranging from mobiles to desktops using The Things Network platform.

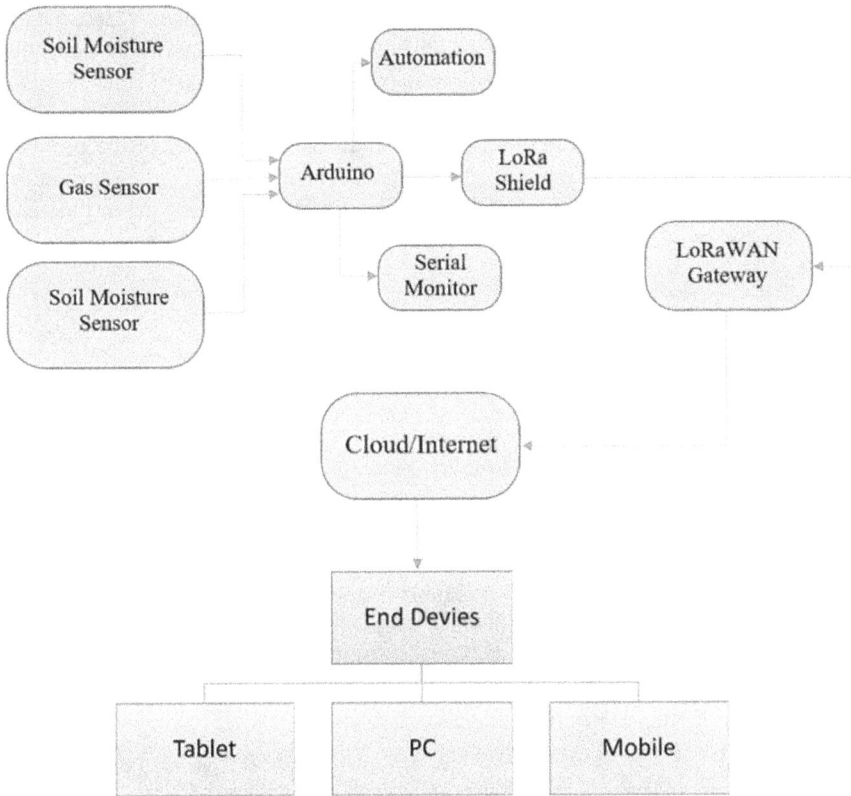

FIGURE 3.1 The complete setup of the sensors, LoRa transmitter, and receiver with the end devices

3.4 SMART AGRICULTURE

Food is the most basic necessity of all living creatures. Humans depend on agriculture for medicinal needs and other raw materials. The rapid increase in world population results in rapid consumption of food and other raw materials worldwide. A rapid escalation in food production to cater to the growing demand is not an easy task [7]. Because of the increase in population and urbanization, there has been a significant decrease in the area of cultivable land. Agriculture has evolved and has tried to keep up with the improvements in technologies which can be termed "The Third Green Revolution". The world is witnessing yet another fundamental modification in the wake of a new industrial revolution that employs the application of modern information and communication technologies into agriculture, in order to deliver sustainable agricultural production. Smart agriculture is the integration of advanced technologies into already existing agricultural practices to increase production quality and efficiency. For the purpose of automation, various parameters such as air quality, water, and soil quality have to be considered. It helps in better decision-making to gain a high-quality output of the product. A technically advanced farming management system aims at boosting productivity based on the inter and intra-field parameters for a significant area [8]. The goal of smart agriculture research is to provide a decision-making support system for agriculture management. The goal of smart agriculture is to maintain the quality of the products while reducing the number and heavy workload of farm workers. Smart agriculture tries to address the issues of population growth, climate change and labor that need a lot of technological attention, from planting and watering of crops to health and harvesting.

3.5   ALGORITHM

Step 1: Start
Step 2: The 10-min timer for the condition checking is started
Step 3: The data is transmitted from the shield to the gateway
Step 4: The data from the gateway is taken to the cloud
Step 5: The data from the cloud can be viewed using an end device
Step 6: After 10 min the condition is checked
    Step 6.1: If the value of the soil moisture sensor is above the threshold or $6.5 > = pH > 7.8$
        Step 6.1.1: Close the solenoid valve
    Step 6.2: If the value is below the threshold value and $6.5 < pH < = 7.8$
        Step 6.2.1: Open the solenoid valve
    Step 6.3: If the gas sensor value is above the threshold
        Step 6.3.1: Alert the user
Step 7: Reset the timer
Step 8: Goto Step 3
Step 9: Stop

3.6   METHODOLOGY

Different crops require different suitable conditions for growth. Therefore, the different limits in each parameter have to be considered in order to automate the process [9]. In this chapter, the air, soil, and water quality are monitored using different sensors interfaced with the LoRa shield. As depicted by the prototype in Figure 3.2, the entire field is sectioned into equal areas and the sensors along with the shield are placed in each section in order to monitor the condition of that section so that appropriate automation measures can be taken. Several such shields, interfaced with different sensors, are placed at various parts of the field. All such shields which are placed within a distance of about 15km from the receiver can be connected to a single gateway. Since the sensors are connected to the shield, the data collected by the sensors are transmitted from the shield to the gateway. The data from the gateway can be taken to the cloud, The Things Network, or other gateways or end devices. Parameters which are obtained in the cloud are being used for the automation of the process.

FIELD SETUP

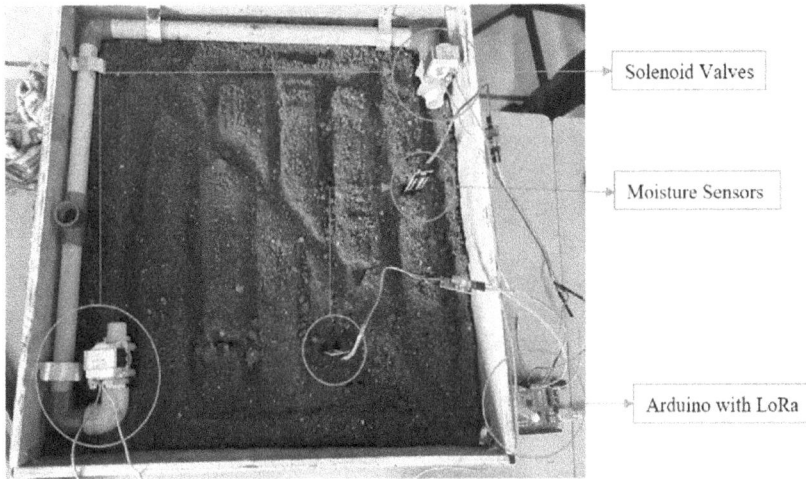

FIGURE 3.2   A prototype of the field setup with the solenoid valve and moisture sensor

3.6.1   Air Quality Monitoring

In order to maximize productivity, suitable air quality has to be monitored. The various gases that are present in the atmosphere include:

1. Nitrogen dioxide
2. Ozone
3. Particulate matters
4. Carbon monoxide
5. Sulfur gas

Of these, in this chapter, the pollutants that are measured are carbon monoxide and sulfide gas. Carbon gas is measured using MQ 9 (Figure 3.3) and hydrogen sulfide is measured using MQ 135 sensor.

3.6.2   Soil and Water Quality Monitoring

Each plant will require different soil nutrients and moisture levels for growth [10]. So, monitoring the various nutrients present in the soil becomes essential. The main parameter which has a major impact on plants is the pH level and the temperature of the water and the soil. Figure 3.4 shows the required sensors interfaced with the shield for the purpose of collecting the data on the various parameters. When the pH is greater than 7 the water is basic. If the pH increases further it implies the water contains a high content of salt and hence making it unsuitable for agriculture. When the pH is less than 7 then the water is acidic and the soil will be unfit for the growth of any plant. Soil and water having a pH in the range of 6.7 to 7.8 is the ideal range. The pH is measured using a pH 4502c sensor.

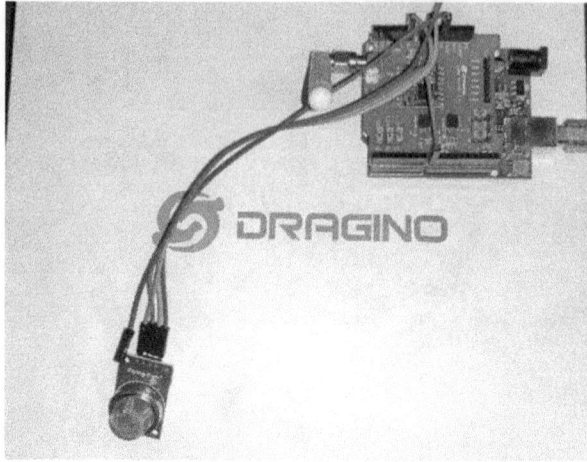

FIGURE 3.3 Connection of an MQ 9 sensor used for air quality monitoring

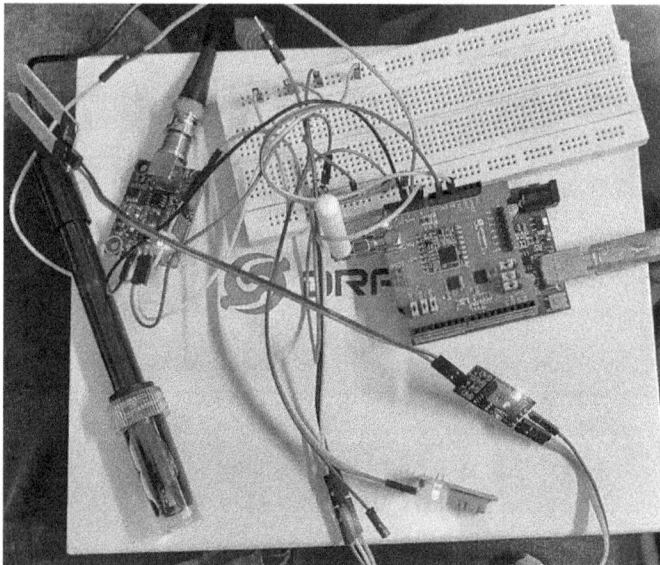

FIGURE 3.4 The various sensors required for soil and water quality monitoring connected with the LoRa shield

The moisture level of the soil is the second most important parameter. Some plants such as rice will require high moisture content at all times, whereas less moisture is sufficient for some plants. So, depending on the type of plant the moisture is measured using YL-69 – YL-38 can be used to determine if the soil is suitable for the growth of that particular plant [11].

3.6.3 AUTOMATION

All the sensors mentioned above are connected with the LoRa shield and gateway and the data is taken to The Things Network. The values are measured continuously. These measured values are compared against a set condition depending on the plant. The conditions are checked every 10 min and depending on the results the corresponding actions are taken by automation.

For example, the moisture sensors placed on the field continuously transmit data to the cloud. When the moisture level is less than a threshold value, in this case 1000, the soil is moist and watering is not required. When the moisture level value is above the threshold value, then the soil requires watering. The value along with the status will be displayed in the cloud.

Based on these values the solenoid valve, which is an electrically controlled device (Figure 3.5), that is placed in each section where the sensors are placed, is turned ON or OFF, allowing water flow to areas which require watering. Thus, automating the watering process (Figure 3.6).

LED 1 : Sensor 1

ON – Watering is Required

OFF – Watering is not required

LED 2 : Sensor 2

ON – Watering is Required

OFF – Watering is not required

Similarly, the other data that are collected can be used for automation thereby increasing crop productivity and also reducing the manual labor required.

3.7 RESULTS AND CONCLUSION

As mentioned earlier, for water and soil quality monitoring the parameters that are measured are pH, temperature, and moisture. In the case of temperature, except for very high variation, there won't be much impact on the plant, so approximate temperature measured using DS18B20 in terms of celsius is sufficient. The pH sensor must be calibrated by testing with different solutions and then placed on the field.

FIGURE 3.5   The solenoid valve used for the purpose of automation

FIGURE 3.6 The status of the valve is indicated with the help of LEDs

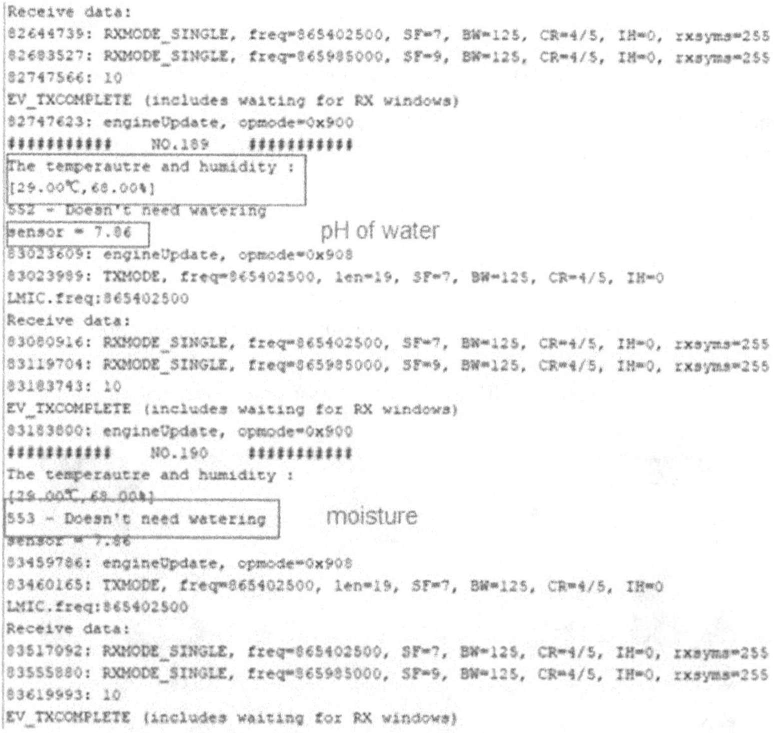

FIGURE 3.7 The pH and soil moisture values received by the Arduino are shown in the serial monitor

These data are encrypted and then taken to the cloud where the decryption is achieved by using a payload function. This decrypted data is then made visible to the end user.

For all the sensor's output values, a threshold value is set as a reference. When there is a deviation in the obtained values from the output of the sensors beyond the threshold, corresponding actions are taken for further use.

For the pH sensor, the pH of the water is around the value of 7.86. Water is a neutral solution and its pH has to maintain in the range of 6.5 to 8. When the value is less than 6.5 it means the solution is acidic and when the value is greater than 8 the water is basic and in either case, it becomes unfit. The serial monitor view of these values is shown in Figure 3.7. In either case, the pH value of water informs that the water is unusable and is monitored regularly, any deviation even of the slightest margin must be reported for necessary actions.

The threshold value for the moisture sensor is set as 1000. If the moisture value is less than 1000, it is noted as the soil has enough moisture and informs the controller that there is "NO NEED FOR WATERING" and if it is greater than 1000, it is noted as dry soil and an instruction is given to the controller to open the solenoid valve in order to "WATER" the region.

In the case of the temperature sensor and the humidity sensor, the threshold values are set depending on the use case of the application or the type of plant to be monitored constantly. All these data are taken to the cloud, Figure 3.8, so it is possible for remote viewing and controlling of the setup (Figure 3.9).

Smart agriculture, air, and water quality management are achieved through LoRa in efficient ways. It is clear that LoRa wireless technology is going to play a big role in the IoT market. Interconnecting devices to create smart cities and industrial and

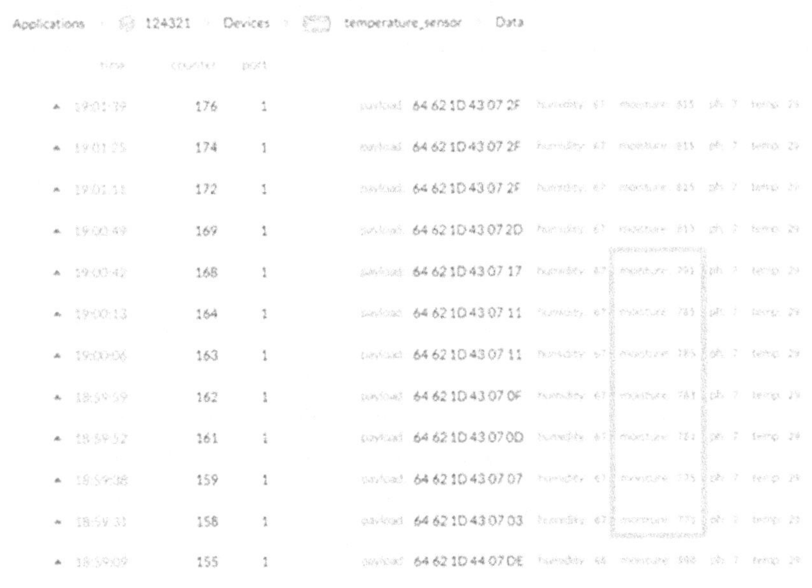

FIGURE 3.8 The various sensor values received in the cloud as viewed from an end device

payload: **64 62 10 20 FF 4F**   MOISTURE1: 1023   MOISTURE2: 847   Status1: "NEED WATERING"   Status2: "NO NEED WATERING"

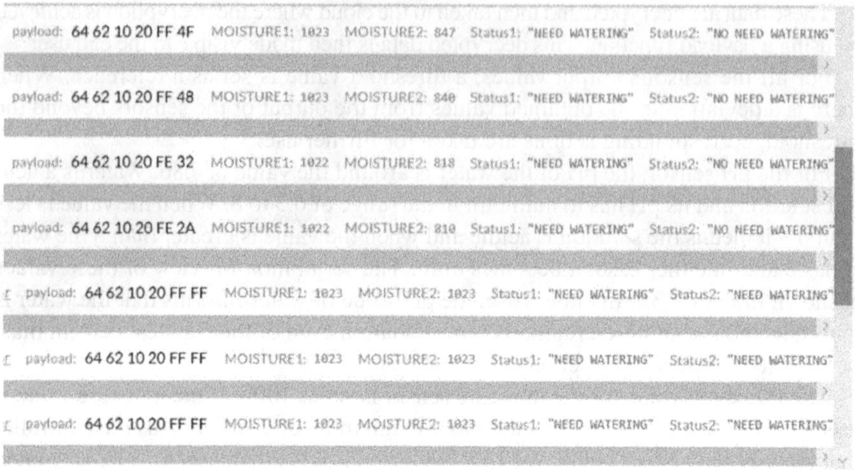

payload: **64 62 10 20 FF 48**   MOISTURE1: 1023   MOISTURE2: 840   Status1: "NEED WATERING"   Status2: "NO NEED WATERING"

payload: **64 62 10 20 FE 32**   MOISTURE1: 1022   MOISTURE2: 818   Status1: "NEED WATERING"   Status2: "NO NEED WATERING"

payload: **64 62 10 20 FE 2A**   MOISTURE1: 1022   MOISTURE2: 810   Status1: "NEED WATERING"   Status2: "NO NEED WATERING"

payload: **64 62 10 20 FF FF**   MOISTURE1: 1023   MOISTURE2: 1023   Status1: "NEED WATERING"   Status2: "NEED WATERING"

payload: **64 62 10 20 FF FF**   MOISTURE1: 1023   MOISTURE2: 1023   Status1: "NEED WATERING"   Status2: "NEED WATERING"

payload: **64 62 10 20 FF FF**   MOISTURE1: 1023   MOISTURE2: 1023   Status1: "NEED WATERING"   Status2: "NEED WATERING"

FIGURE 3.9   The value along with the status of all the sensors is displayed using The Things Network Platform

commercial solutions, while reducing the limitations of other wireless technologies such as power and other overheads.

While having to monitor several areas, placing the various sensors interfaced with the LoRa shield within a 4km radius of the LoRa gateway in an urban area will not only reduce the power consumed but will also reduce the complexity of the system.

The future scope of this project involves allowing the system to make decisions based on the available data. Instead of the developer having to provide the steps to be implemented when each time the threshold value is met, the system by itself is able to predict the future possibility based on the previously available data and take the required measures. This can be achieved through machine learning.

REFERENCES

1. LoRa alliance. https://www.lora-alliance.org/what-islora/technology
2. Lorenzo Vangelista, Andrea Zanella, and Michele Zorzi. 2015. "Long-range IoT technologies: The dawn of LoRa." *Future Access Enablers of Ubiquitous and Intelligent Infrastructures*, 51–58.
3. Semtech. http://www.semtech.com/wireless-rf/lora/LoRa-FAQs.pdf
4. Umber Noreen, Achene Bounceur, and Laurent Clavier. 2017. "A study of LoRa low power and wide area network technology." In *International Conference on Advanced Technologies for Signal and Image Processing (ATSIP)*.
5. Oratile Khutsoane, Bassey Isong, and Adnan M. Abu-Mahfouz. 2017. "IoT devices and applications based on LoRa/LoRaWAN." In *IECON 2017 – 43rd Annual Conference of the IEEE Industrial Electronics Society*, 6107–6112.
6. Aloys Augustin, Jiazi Yi, Thomas Clausen, and William Townsley. 2016. "A study of LoRa: Long range and low power networks for the internet of things." *Sensors*, 16(9), 1466.

7. "Why Smart Agriculture is the need of the hour." September 26, 2018. Accessed on: December 9, 2019 [Online]. Available: http://www.isaf-forum.com/related-news-details .html

8. S. Jagannathan, and R. Priyatharshini. 2015. "Smart farming system using sensors for agricultural task automation." In *2015 IEEE Technological Innovation in ICT for Agriculture and Rural Development (TIAR)*, 49–53.

9. Nattapol Kaewmard, and Saiyan Saiyod. 2014. "Sensor data collection and irrigation control on vegetable crop using smart phone and wireless sensor networks for smart farm." *IEEE Conference on Wireless sensors (ICWiSE)*, 106–112. DOI: 10.1109/ICWISE.2014.7042670

10. Soil and water quality, an agenda for agriculture. http://www.nap.edu/read/2132/chapter/6

11. Weng YeowTan, Yi LungThen, Yao LongLew, and Fei SiangTay. 2019. "Newly calibrated analytical models for soil moisture content and pH value by low-cost YL-69 hygrometer sensor." *Measurement*, 134, 166–178.

4 Local Binary Pattern-Based Criminal Identification System

*Dhana Srinithi Srinivasan, Soundarya Ravichandran,
Thamizhi Shanmugam Indrani, and G. R. Karpagam*

CONTENTS

DOI: 10.1201/9781003307716-4

4.1   INTRODUCTION

India is known for its various festivities and celebrations which take place during different seasons of the year. Lawbreakers ardently wait for opportunities to commit felonies, especially during such seasons and many innocent people fall prey to them. The discipline of criminal identification has gained importance due to the increasing crime rate in the past few decades. These situations expect efficient identification methods for immediate actions.

T Nagar, Chennai, is famous for the variety of items that one can buy there, the cheap prices of products and the large shops which provide large discounts during festival seasons. These factors make it a favorite destination for middle-class shoppers. During festival seasons, the streets and by-lanes of T. Nagar are usually filled with a sea of people, shopping for gifts, clothes, and jewelry, ahead of the festival. Police statistics state that, during such seasons, the crowd usually peaks after 6.30 pm. About 500,000 people throng Ranganathan Street, Pondy Bazaar, and Panagal Park, the main shopping areas of T Nagar. The number increases to about 800,000 by 6.30 pm. Police on-duty stated that they usually expect at least 1,000,000 people by the time the shops down their shutters at around 11:00 pm. Shops that usually close by 10:00 pm, keep their doors open till late at night to accommodate the shoppers.

In 2018, ahead of the Diwali shopping season, several police personnel on patrol duty were given jackets that had inbuilt cameras for better surveillance, In addition to this, 750 CCTV cameras were installed at several locations and a few drones with cameras were left to hover around the area. Special cameras with capabilities to zoom and view people and the items they shopped for were also installed at specific strategic points. Such a surveillance system will give rise to a number of facial images of shoppers, along with images of the items they purchase and their activities during the course of time spent there. Such data can not only be used to easily detect criminal and fraudulent activities, but also to identify the lawbreakers behind such activities. This data can be matched with an already existing database containing facial images of criminals and can be sent to the nearest police station based on the locality of the camera that captured the image and immediate measures could be taken. The database of criminal facial images is updated periodically by the concerned officials or organizations as new felons are identified. This allows the system to work with an up-to-date set of data.

4.2   TERMS AND TERMINOLOGIES

4.2.1   Detection

Face detection is the first and essential step for face recognition, and it implies an algorithm to detect faces in the images. This method must be efficient and robust. Detection systems use an algorithm that tends to be simpler and is efficient in discriminating between face and non-face images or face and non-face portions in a single image.

4.2.2 PREPARATION

In the criminal identification system, the images should be named in such a way that it is easy to navigate through the directories and folders to extract the training and test image files. Once the navigation to the right folder is done and the image is accessed, it is resized and passed for processing.

4.2.3 PREDICTION

Once the training is done, the trained recognizer is used for predicting the class label of the test image. Here, the timestamps before and after prediction can be used to rate the efficiency of the system. Confidence measurement is also used to predict the class label.

4.2.4 CONFIDENCE

Confidence is a measure of the distance between the test image and the training images associated with the particular label. So, the lesser the confidence, the better the similarity.

4.2.5 LOCAL BINARY PATTERNS (LBP)

This is the concept used behind the detection system. The original LBP operator labels the pixels of an image by thresholding the 3-by-3 neighborhood of each pixel with the center pixel value and considering the result as a binary number. Local primitives which are codified by the 256-bin histogram of the labels computed over an image include different types of curved edges, spots, and flat areas. The most important properties of LBP features are their tolerance against monotonic illumination changes and their computational simplicity.

4.2.6 SUMMARIZATION

When the criminal label has been identified, the corresponding criminal record needs to be extracted from the database. The history of the criminal would be helpful for the officials for further moves in identification. A concise and read-easy summary of the report would be more useful.

4.2.7 TEXTRANK

TextRank is a graph-based ranking model for text processing. The blank spaces, new lines, tab spaces, and stop words are removed from the text. A similarity matrix is generated to indicate the similarity among the various sentences. A similarity graph is constructed based on the similarity matrix; the sentences are then ranked based on the graph.

4.3   RELATED WORKS

A literature survey helps to discover what statistical knowledge exists related to the research topic. It helps to find gaps in existing research so that new original ideas can be generated. The relevance of the proposed idea can be justified. Few references are made and the content of the paper works are summarized below.

Reference [1] provides an insight into the role software systems are playing in the Police Forces. It categorized crimes as major and volume. The former included high-profile crimes such as murder, armed robbery, etc. The latter included burglary and shoplifting. Various methods of identification for respective categories of crime have been stated. It gives a view that different types of relational database management systems for recording and subsequent analysis of crime are used. A face recognition system typically compares a test image with a database of criminal facial images and checks if a match exists. Various techniques are available for this purpose.

Reference [2] proposes a near real-time face recognition system that projects facial images over a "face space". This face space is defined by the eigenvectors, which correspond to the entire face, as a whole, and does not pertain to individual features within a face. The limitation of this system is that any noise or occlusion in an image degrades the system's performance significantly.

Reference [3] proposes a system that uses a principal component analysis in which every image is represented as an eigenface, which is a linear combination of weighted eigenvectors. These eigenvectors are obtained from the covariance matrix of a training image set and the weights are found after selecting a set of the most relevant eigenfaces. Recognition is performed by projecting a test image onto the subspace spanned by the eigenfaces and then classification is done by measuring the minimum Euclidean distance. This system requires that the images be normalized and have a frontal view. Fixing a threshold to filter out the non-face images is done manually.

Reference [4] proposes a system that uses Linear Discriminant Analysis (LDA) for face recognition. LDA finds the most discriminant projection vectors which can map high-dimensional samples onto a low-dimensional space. This system is an improved version of the traditional LDA that addresses the small sample size problem. The above-mentioned approaches focus on providing descriptions for the entire image and not based on local features of the face. These techniques are typical examples of holistic image-matching methods.

On the other hand, Multi-Dimensional Scaling (MDS) [5], Fast Fourier Transforms (FFTs) [6], Local Binary Patterns (LBPs) [7], Two-Dimensional Maximum Local Variations (2DMLV) [8] are examples of feature-based methods. Feature-based approaches process the images by extracting the various local features of the image, such as eyes, nose, mouth, and other prominent marks. Once it locates the positions of such features, it computes the relationships (geometric) among them and generates a feature vector for each image.

The advantage of feature-based methods over holistic approaches is that feature-based methods are more immune to local changes such as expression, occlusion, and misalignment.

Reference [7] proposes a feature-based face recognition approach using LBPs. A face in an image is split into non-overlapping regions. An LBP histogram is used to describe each of the regions. The histograms are concatenated to form a feature histogram that represents the entire face. This system was found to be robust and allowed very fast feature extraction.

Reference [9] proposes a face recognition system that is implemented using a variation of the LBP algorithm that considers different neighborhoods and more pixels to compute the feature vectors. This improves the system's robustness in handling noise and different illumination conditions.

Text summarization involves reducing the length of a long piece of text, in order to generate a precise, concise, and meaningful version thereby improving the text's readability. There are two major approaches to text summarization: extractive and abstractive methods.

TextRank [10] is an extractive text summarization technique proposed by Rada Mihalcea and Paul Tarau where text in articles is split into individual sentences. Each sentence is then described using a vector, after which similarities between sentence vectors are calculated and stored in a matrix. This matrix is converted into a similarity graph, with sentences as vertices and similarity scores as edges. The sentences with the highest ranks are then picked to form a summary of the original text.

LexRank [11] is another extractive text summarization where every sentence represents a node, and the edges represent similar relationships between sentences. The similarity between sentences is computed by the frequency of word occurrence in a sentence using the IDF-modified cosine formula of the TF-IDF formulation. This similarity measure is then used to build a similarity matrix and consequently a similarity graph between sentences. A threshold value is used to filter out the relationships between sentences whose weights fall below the threshold. The node with the highest degree is produced as a result.

Reference [12] proposes an automated facial recognition system, for a criminal database, which uses the principal component analysis approach. This system was proposed for use primarily in Malaysia, where the existing criminal identification system relied upon criminal thumbprints. This study lacked provisions for criminal record summarization.

Based on the conclusions drawn from the literature survey, software tools and algorithms for the different phases of the system have been selected. The major considerations for the selection process include ease in usability and efficiency.

The categories and methodologies highlighted in Figure 4.1 are taken into consideration.

Among the various methods with which one can build a criminal identification system, face recognition was chosen due to its prediction accuracy and efficiency. The approach chosen for face recognition is Machine Learning wherein, the system is trained by giving various images of a particular face as input and is then made to identify the label given a new input image.

The language chosen is Python due to its easy installation and usage of libraries. Python's OpenCV library is used for the implementation of the face recognition

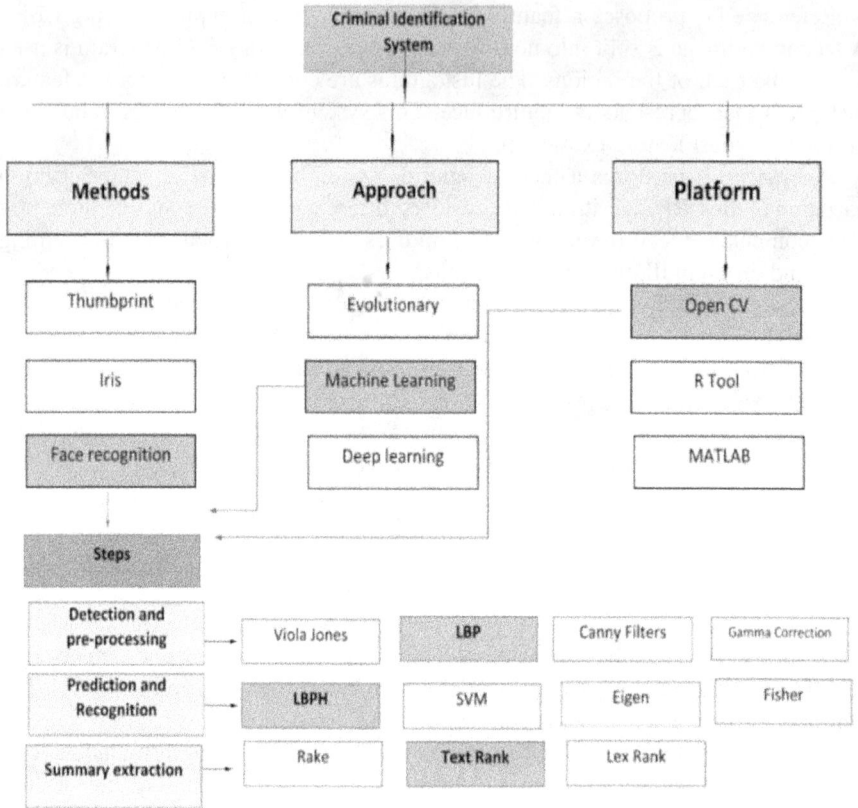

FIGURE 4.1   Context of research

system. OpenCV supports the use of various frameworks and provides different programming functions for real-time computer vision.

The detection mechanism chosen is LBPs. It allows for easier pre-processing and produces better results under different lighting conditions. LBP histogram is used for prediction, its working is similar to that of LBP.

The TextRank algorithm is used for criminal record summarization. It is an extractive text summarization technique where ranks are associated with each sentence and the sentence with the maximum rank effectively summarizes the entire record.

4.4   CONCEPTUAL ARCHITECTURE

The architecture adapted in this system is a layered architecture. Each of the components within the system is organized into horizontal layers, with each layer performing a specific function within the application. Each layer has a specific responsibility which contributes to the proper functioning of the system. The perception layer, physical layer, processing layer and service layer interact with each other in an orderly fashion and form their own abstraction around the work that needs to be done in order to satisfy a request. Layered architecture improves the maintainability, scalability and flexibility of the entire system as each layer is concerned with only a particular task and thereby makes a contribution to satisfying a request.

4.4.1 PERCEPTION LAYER

The perception level layer emphasizes live input from surveillance cameras, which satisfies the IoT-based image acquisition portion of the module, and the output is achieved by producing a web-based report containing the criminal's label and summary of his/her report. The input image is mediated to the physical layer, hence satisfying the communication portion of the module. Once the backend process of detection and prediction has been made, the predicted image, class label and extracted summary for the corresponding detected criminal are displayed using suitable devices such as wireless handsets/PCs so that immediate actions could be taken.

4.4.2 PHYSICAL LAYER

In the criminal identification system, all the training images must be stored in the cloud repository such that they are easily accessible by the modules that require them. Similarly, the input images should also be stored by following a particular convention, as required by the other algorithms. The OS library is used to navigate through the directories and folders to extract the training and input image files. For this application, training images are stored in a folder named "training-data". This folder contains a sub-folder for every criminal image. Similarly, test images are stored in a folder named "test-data". Once navigation to the correct folder is done and the image is accessed, it is resized and passed for processing. The criminal records need to be stored in the cloud repository prioritized by the severity of the criminal act.

4.4.3 PROCESSING LEVEL LAYER

4.4.3.1 LBP-Based Face Detector

Face detection is the first and most essential step for face recognition, and it is used to detect faces in images. Since criminal identification systems need to be flexible enough to be executed on mobile products, such as handheld computers and mobile phones, the face detection method must be efficient and robust. Face detection systems that use LBP-classifiers tend to be simpler and are efficient in discriminating face and non-face images or face and non-face portions in a single image. The original LBP operator labels the pixels of an image by thresholding the 3-by-3 neighborhood of each pixel with the center pixel value and considering the result as a binary number. Local primitives which are codified by the 256-bin histogram of the labels computed over an image include different types of curved edges, spots, and flat areas. The most important properties of LBP features are their tolerance against monotonic illumination changes and their computational simplicity. To consider the shape information of faces, they divide facial images into small non-overlapping regions. The LBP histograms extracted from each sub-region are then concatenated into a single, spatially enhanced feature histogram. The extracted feature histogram describes the local texture and global shape of face images. The area within the image beside the facial portion is considered to be the non-facial portion and is ignored.

4.4.3.2 LBPH-Based Face Label Predictor and Face Recognizer

The LBPH face recognizer is used in this system for the recognition of faces. The recognizer is first trained with a collection of "faces" and "labels". Once the training is done, the trained recognizer is used to predict the class label of the test image. The predictor uses the LBPH methodology specified in the detection module for predicting the test image's class label. Here, the timestamps before and after prediction can be used to rate the efficiency of the system. The confidence associated with the prediction can also be used for this purpose. Confidence is a measure of the distance between the test image and the trained images associated with the label. So, the less confidence, the better the similarity.

4.4.3.3 TextRank-Based Summary Extractor

The last phase of CIS is to provide the user with a concise and easy-to-read summary of the report associated with the predicted criminal. TextRank, a graph-based ranking model for text processing. This is a derivation of the PageRank algorithm used to rank web pages. As mandatory for almost all Machine Learning algorithms, the text must be pre-processed. The blank spaces, new lines, and tab spaces are removed from the text. In addition to this, the stop words are also removed from the lines. Once this is done, a similarity matrix is generated to indicate the similarity among the various sentences. A similarity graph is constructed based on the similarity matrix; the sentences are then ranked based on the graph. These ranks are stored in a list and are sorted. The ranks of sentences that top the list are selected and displayed. These sentences, when put together, result in a concise summary.

4.4.4 CRIMINAL IDENTIFICATION AS A SERVICE

The input from the perception layer is forwarded to the cloud that provides criminal identification as a service. This layer uses three fundamental operations: publish, find, and bind. Service providers (SP1: LBP detector, SP1: Predictor, SP1: Recognizer, SP2: TextRank summarizer) publish services to a service registry. Service requesters find required services using a service registry and bind to them. These ideas are shown in Figure 4.2 in relation to the criminal identification system.

4.5 PROTOTYPE IMPLEMENTATION

The prototype is implemented to check for the feasibility of the design. The actual flow of the system is made transparent so that we could discern the complete purpose

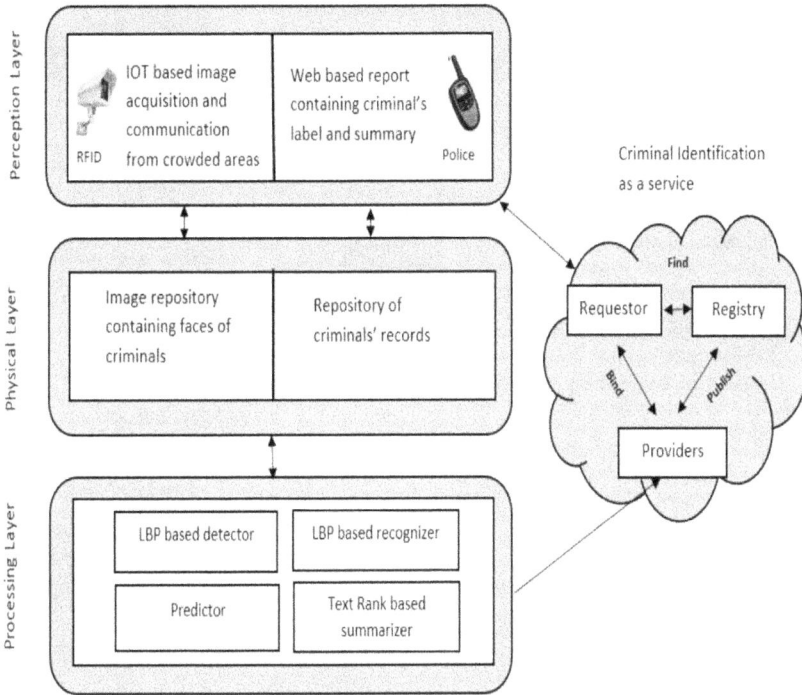

FIGURE 4.2 Conceptual architecture for the criminal identification system

and structure of the system. It can improve the quality of requirements and specifications so that the early determination of what the end user really wants can be made which can result in faster and less expensive software. The flow of the identification system using face detection for various modules along with the pseudocode is described below.

4.5.1 DETECTION MODULE

Step 1: The test image is given as an input
Step 2: The image is converted into a greyscale
Step 3: The detector is loaded and faces, if any, are detected
Step 4: The detected portion is extracted and stored

4.5.2 PRE-PROCESSING MODULE

Step 1: Training data is found through navigation
Step 2: Images are resized appropriately for prediction
Step 3: The detection module is called
Step 4: Returned labels are appended

4.5.3   PREDICTION MODULE

Step 1: Recognizer is launched and trained
Step 2: A test image is passed to the detector
Step 3: A class label for the test image is predicted
Step 4: Confidence is accessed, and the timestamp is noted

4.5.4   SUMMARIZATION MODULE

Step 1: The text is cleaned by removing blank spaces and stop words
Step 2: A similarity matrix and a graph are generated for the text
Step 3: Sentences are ranked based on frequency and stored
Step 4: Top-ranked sentences are selected

4.6   ASSESSMENT

Testing the system is an essential phase since it makes sure of the reliability and quality of the application. Testing is required for the effective performance of the system. It's important to ensure that the application should not result in any failures because it can be very expensive in the future or in the later stages of development. Assessing the system is important to evaluate the procedure in which results are obtained in terms of time, cost, and relevance to the problem. A confusion matrix is used for evaluation. A confusion matrix is a specific table layout that allows visualization of the performance of a supervised learning algorithm. The criminal identification system belongs to supervised learning since the training data of criminal faces and their records are used for training the classifier and the input test image is predicted by the trained classifier. Each row of the matrix represents the instances in a predicted class while each column represents the instances in an actual class or vice versa.

True Positive (TP): Observation is positive and is predicted to be positive
False Negative (FN): Observation is positive but is predicted to be negative
True Negative (TN): Observation is negative and is predicted to be negative
False Positive (FP): Observation is negative but is predicted to be positive
Condition Positive (P): The number of real positive cases in the data
Condition Negative (N): The number of real negative cases in the data

The results from the tried experiment are mentioned in Figure 4.3 in the form of a confusion matrix and the performance measure for the classifier is plotted in Figure 4.4.

Accuracy = (TP+TN)/Total
Precision = TP/(TP+FP)
Recall = TP/(TP+FN)
f-measure =2×(Recall×Precision)/(Recall+Precision)

Truth data

	Class 1	Class 2	Class 3	Class 4	Classification overall	Producer Accuracy (Precision)
Class 1	14	1	0	0	15	93.333%
Class 2	0	22	0	2	24	91.667%
Class 3	0	0	18	0	18	100%
Class 4	0	0	0	13	13	100%
Truth overall	14	23	18	15	70	
User Accuracy (Recall)	100%	95.652%	100%	86.667%		

Classifier results

FIGURE 4.3 Confusion matrix for face detection

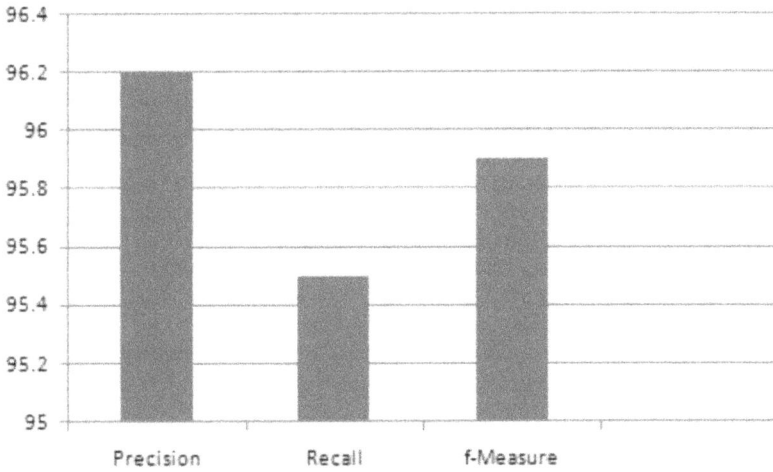

FIGURE 4.4 Plot of precision, recall, and f-measure

4.7 DISCUSSION AND CONCLUSION

The destined purpose has been achieved by first efficiently recognizing a criminal's face among the crowds. This has been achieved by the LBP algorithm to detect each face iteratively from multiple images. As soon as a face matches the criminal images record, information is sent to the identification department along with a short summary of the criminal's fraudulent history. The performance has been improved by the selection of efficient face detection, recognition, classification, and text summarization algorithms.

Additional features that can be incorporated into this project include geographical tracking of the criminal and reporting to the identification department. Algorithms for face detection such as Viola–Jones, Canny filters, and Gamma correction can be used as alternatives. Similarly, alternative classification algorithms include support vector machines, Eigen and Fisher prediction and recognition algorithms can also be used. For the non-functional aspect, security would be a major consideration and the Open Web Application Security Project (OWASP) can be employed to remove security vulnerabilities if any.

REFERENCES

1. Richard Adderley, and Peter Musgrove (2001), "Police Crime Recording and Investigation Systems – A User's View", *Policing an International Journal of Police Strategies and Management*, 24(1), 100–114.
2. M. A. Turk, and A. P. Pentland (1991), "Face Recognition Using Eigenfaces", *IEEE Computer Society Conference on Computer Vision and Pattern Recognition*, pp. 586–591.
3. Güneş Erkan, and Dragomir R. Radev (2004), "LexRank: Graph-Based Lexical Centrality as Salience in Text Summarization", *Journal of Artificial Intelligence Research* 22, 457–479.
4. Kyungnam Kim (1996), "Face Recognition Using Principal Component Analysis", [online]. http://scholar.fju.edu.tw/%E8%AA%B2%E7%A8%8B%E5%A4%A7%E7%B6%B1/upload/058029/content/991/D-7604-07699-E.pdf
5. Muhammad Burhan, Rana Asif Rehman, Bilal Khan, and Byung-Seo Kim (2018), "IoT Elements, Layered Architectures and Security Issues: A Comprehensive Survey", *Sensors (SENSORS-BASEL)*, 18(9). 10.3390/s18092796.
6. Li-Fen Chen, Hong-Yuan Mark Liao, Ming-Tat Ko, Ja-Chen Lin, and Gwo-Jong Yu (1999), "A New LDA-Based Face Recognition System Which Can Solve the Small Sample Size Problem", *Pattern Recognition*, 33, 1713–1726.
7. Timo Ahonen, Abdenour Hadid, and Matti Pietikainen (2006), "Face Description with Local Binary Patterns: Application to Face Recognition", *IEEE Transactions on Pattern Analysis & Machine Intelligence*, 28(12), 2037–2041.
8. Pavel Král, Ladislav Lenc, and Antonín Vrba (2018), "Enhanced Local Binary Patterns for Automatic Face Recognition". *International Conference on Artificial Intelligence and Soft Computing*, Springer, pp. 27–36.
9. Nurul Azma Abdullah, Md. Jamri Saidi, Nurul Hidayah Ab Rahman, Chuah Chai Wen, and Isredza Rahmi A. Hamid (2017), "Face Recognition for Criminal Identification: An Implementation of Principal Component Analysis for Face Recognition", *AIP Conference Proceedings* 1891(1), 020002.
10. Quanxue Gao, Feifei Gao, Hailin Zhang, Xiu-Juan Hao, and Xiao Gang Wang (2013), "Two Dimensional Maximum Local Variation Based on Image Euclidean Distance for Face Recognition", *IEEE Transactions on Image Processing*, 22(10), 3807–3817.
11. Gaurav Jain, Shahebaz Khan, Nikhil Jagtap, and Sandip Gangurde (2015), "Face Recognition by Multi-Dimensional Scaling (MDS)", *International Journal of Modern Trends in Engineering and Research*.
12. Lina Zhao, Wanbao Hu, and Lihong Cui (2012), "Face Recognition Feature Comparison Based SVD and FFT", *Journal of Signal and Information Processing*, 3(2), 259–262.

5 Enhancing Customer Experience in Restaurants Using Augmented Reality-Based Marketing Strategy

S. J. Syed Ali Fathima, N. Yashwanth,
N. Akshaya, and K. G. Sethu Balaji

CONTENTS

5.1 INTRODUCTION

Innovation was a substantial word 15 years ago; however, the new age is all forward-thinking. We could see the effect of innovation becoming extremely significant in everyday life and even more specific about nutrition and diet i.e., nourishment. People live to eat and that is possibly the reason that the Indian nourishment and basic food item market is the world's 6th largest, with retail contributing to 70% of the deals. Innovation is developing exponentially across all industries; Gartner predicts the utilization of 6.4 billion associated gadgets this year, with a 30% expansion year-over-year. That hops to 20.8 billion gadgets by 2020. There is additionally a fear that the quick development of innovation in PCs and robots could be a risk to the occupations of workers. Rather, innovation is an approach to engage industry instead of supplanting the workers with robots. With innovation, individuals would not need to consider the issues which happen so frequently. Instead, they could consider settling the issues which haven't happened but may happen in the future, much the same

DOI: 10.1201/9781003307716-5

as a researcher. There is tremendous potential in innovation showcase; techpreneurs need to fill the void.

5.2 LITERATURE REVIEW

R. Silva et al. presented an outline of essential parts of Augmented Reality (AR) and the significant ideas of this innovation. When structuring an AR framework, three angles must be given top priority: (1) Combination of genuine and virtual universes; (2) Interactivity progressively; (3) Registration in 3D. Five significant classes of AR can be recognized by their showcase type: Optical See-Through, Virtual Retinal Systems, Video See-Through, Monitor-Based AR, and Projector-Based AR. Optical See-Through AR utilizes a straightforward Head Mounted Display (HMD) to show the virtual condition over the present reality. The Virtual Retina Display (VRD) gives a balanced light emission from an electronic source onto the retina of the eye delivering a rasterized picture. Video See-Through AR utilizes a misty HMD to show blended video of the Virtual Environment (VE) and the view from cameras on the HMD. Screen-Based AR likewise uses consolidated video streams yet the showcase is an increasingly traditional work area screen or a handheld presentation. Projector-Based AR utilizes real world entities as the projection surface for the virtual condition. Augmented Reality innovation has numerous potential applications in a wide scope of fields, including simulation, instruction, prescription, designing and assembling [1].

Jeff et al. proposed a versatile application that facilitates a way to make 3D models, which targets facilitating and pulling in understudies to become familiar with the 3D displaying ideas and abilities. There are some markerless AR applications, in which the 3D scene is exclusively distinguished by PC vision strategies without the guide of printed markers. These strategies are helpful and ready to be applied to enlarge a bigger scene. A cell phone is much the same as a compact camcorder, which empowers individuals to make 3D models whenever and anyplace. The proposed application is actualized over the condition of-workmanship innovations. They utilized the Vuforia Augmented Reality toolbox for AR marker acknowledgment, and OpenGL ES and Unity3D (which are proficient 3D systems) for rendering the 3D models. The client needs to move the printed AR markers to the ideal 3D position by hand. The camcorder of the cell phone keeps track of the cutting-tangle marker and other AR markers being used and afterward renders the compared 3D model on the screen. The client studies indicated that the clients felt the UI is more normal than conventional work area 3D displaying applications. They have distinguished a few perspectives that help individuals to display 3D objects. For instance, visual intimations are significant for them to evaluate the separations in each measurement. In addition, clients recommended auto-arrangement would make the 3D displaying progressively productive [2].

To portray a blend of advancements, Andre Lemieux empowered a constant blending of PCs produced content with a live video show. This innovation has its underlying foundations in the field of software engineering interface results. AR frameworks are based on three significant structure squares, following and enlistment, show innovation, and real-time rendering. It likewise examines the computer-generated experience. Virtual is characterized as being simulated or imaginary and the truth is characterized to be something that is genuine or actual. AR standards and a couple of advances must be joined to make this adoptable: worldwide following technologies, remote correspondence, area-based figuring, and administrations and wearable computing. The paper talks about the utilization of AR in various fields such as medicine, military, visualization, robotics, geospatial, urban arranging, and structural designing. In medicine it is utilized to envision the X-beam, X-ray in 3-D view and ultrasound imaging [3].

Mauricio Hincapié proposed Augmented Reality with applications in aeronautical maintenance. The paper presents instances of Augmented Reality applications and shows the plausibility of Augmented Reality arrangements to support undertakings, underlining focal points it could present. AR enables the client to see this present reality, expanding it with superimposed virtual articles. As such, while VR replaces reality, AR supplements it, making a domain in which genuine and virtual articles pleasingly coincide. AR misuses clients' perceptual-engine aptitudes in reality, making an extraordinary kind of human–machine association. In the AR application for support of a plane fuel channel, we can see how, superimposed to the image of genuine articles, some virtual items are drawn. Most basic AR arrangements utilize only one camera, yet multiple cameras can be utilized if necessary. At that point, to coordinate the virtual and this present reality, objects developments in the two universes must be followed. While in Virtual Reality (VR) everything is misleadingly made, so articles' positions are characterized in AR the framework needs to pursue changes in the real world and correspondingly adjust the virtual world to successfully coordinate the reality view. AR innovation is amazingly adaptable and, especially in the upkeep industry, it may very well be effectively executed in a few processes. From an efficient perspective, businesses can utilize AR to bring down procedures' operational expenses and, in this manner, support their development and advancement: preparing specific laborers is costly in any industry [4].

ROMA interactive fabrication system with Augmented Reality (AR) and a robotic 3D printer lets users integrate real-world constraints into a design rapidly. This allows them to create well-proportioned tangible artifacts or to extend existing objects. The designer wears the AR headset and designs in print volume using AR controllers and several systems explore the interaction aspect of digital modeling [5].

Jishuo Yang explored the idea of existing enlarged reality executions to create a product and equipment approach to actualize an increased reality programming interface. This venture centers around the development of a programming interface that permits the showcase and communication with a 3D model of a human body. The paper discussed existing arrangements such as ARToolKit and its subsidiaries, ARTag, etc. The result shows one reason for the absence of computerized tabletop AR may be the poorly designed area of the tabletop's cameras. The utilization of AR on a tabletop for the most part requires a versatile invisible/ hidden camera, which isn't found on customary tabletop surfaces. Other confinements of 3D models utilized for portable usage are that they are not quick enough and need more memory to conquer its multifaceted nature diminished to accomplish a worthy memory footprint. As a large portion of the current AR libraries are produced for PC, they do include advancements that would have made ongoing AR conceivable on a mobile. So, the proposed framework shows an evidence that adequately incredible gadgets will land sooner rather than later [6].

Analysts have started to address issues in showing data in AR, brought about by the idea of AR innovation or presentations. Work has been done to remedy blunders and abstain from concealing basic information because of depth issues. A key necessity for improving the rendering nature of virtual realities in AR applications is the capacity to naturally catch the ecological light data. Application advancements can benefit from outside assistance by utilizing accessible libraries. One of them is ARToolKit, which gives PC vision procedures to ascertain a camera's position and direction compared with stamped cards so virtual 3D articles can be overlaid unequivocally on the markers [1]. Augmented Reality is enhancing the view of reality by supplementing virtual objects using technology. AR applies to all senses such as the sensing of light, hearing, smell, and touch, and incorporating specialized processors, sensors and I/O devices. The recent developments are head-mounted displays, handheld displays, and spatial augmented reality. AR visualize the real-world input device and superimposes the image or videos in the real world. Its working is based on the position of markers and identifying the location of the device through GPS. The system will mimic the human brain during further development [7].

Nor Farhah Saidin proposed Augmented Reality in education, and they present that innovation has been inserted in training and the outcomes show a positive effect on learning and instructing styles. The investigation of science is a complex process that incorporates distinguishing an issue, examining the issue, making speculations, arranging the information, testing the speculations, gathering the information, and reviewing the results. Instances of representation advancements that have been analyzed in past research incorporate movement, virtual conditions, and re-enactment. Dede et al (1996) recommend that research studies can improve the dominance of AR conceptual ideas of using virtual situations intended for training environments. AR is another approach to improve the learning of 3D shapes of the conventional strategy wherein instructors utilize wooden objects. The technology to build the AR device with the technique of sending the picture, perceiving the content, and afterward getting the importance of the content is quite tedious. This is on the grounds that the innovation utilized the 3G system to associate with the Internet. However, the utilization of this sort of innovation is developing gradually in Malaysia, particularly in the instruction field. In this way, more scientists in the instruction field ought to explore the capability of AR to improve the educating techniques in the nation's instruction framework and to improve the productivity of the educating and learning process [8].

5.3　PURPOSES AND OBJECTIVES

Restaurants certainly have a lot of time to cooperate with their clients. In step with the investigation, the usual wait time per client is 23 min, but almost a third of the group is waiting for up to 30 min. Because of the issues in preparing food, long wait times, challenges in persuading the new choice of dishes the rise of expanded reality in the nourishment and, the interest in app development companies is also expanding. At the time clients request dishes, they don't have any administration over enduring time and check dishes upheld servers' portrayals or confirm the web to investigate a spread of dishes.

The objective of the system is to shape menu cards with a 3D perception of items and AR narrating on the table. It is superior to the existing arrangement since it will be easy to understand, offers unmistakable comprehension to the Indian customer and tweak innovation-based arrangements in giving/focusing on the information required utilizing old systems. AR technique to produce this drifting innovative idea on the menu card to visualize food items will have great impact on customer experience.

The advantages of AR apps are quite exceptional. During this technology era, it has become essential for each business to adopt the most recent technology trends. AR is one of the foremost developing technologies that restaurants must integrate to expand and stand out from the competition.

This chapter proposes the use of the following techniques to promote the marketing of the restaurants.

• Live visualization by integrating Snapchat 3D banner ads
• AR storytelling on the table
• AR food on social media
• Web AR experience
• 3D visualization for online food ordering and delivery companies

5.4 PURPOSES AND OBJECTIVES

Augmented Reality is used to describe a combination of technologies that enable the real-time mixing of computer-generated content with a live video display. This technology has its roots in the field of computer science. This chapter presents the idea of an AR-based 3D visualization of menu card items by scanning a QR code and using image augmentation. This chapter suggests Vuforia for model development, AWS for storage, and Unity for the projection of the AR models. Unity 3D is a software game engine that helps to develop Augmented Reality applications. The workflow of the AR-based 3D visualization of images is shown in Figure 5.1.

In this chapter, we present menu cards with a 3D representation of items utilizing a QR code and augmented pictures dependent on the QR code. The framework can extricate the data implanted in a QR code and show the data in an all-inclusive 3D structure with the QR code being the customary AR marker. Customary AR frameworks regularly utilize a planned example (the marker) to recoup the 3D scene structure and distinguish the article to be shown on the screen. In these frameworks, the marker is utilized uniquely for following and ID. They don't pass any other data. The QR code has the benefit of a large data limit and is like an AR marker in appearance. Along these lines, additional fascinating and helpful applications can be created by

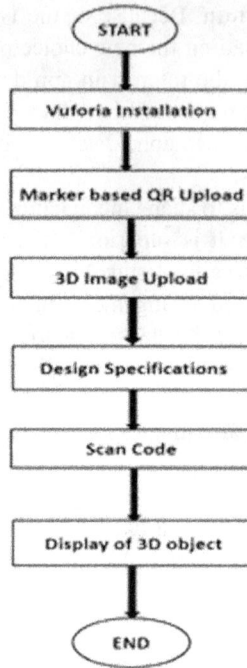

FIGURE 5.1　Workflow of AR-based 3D visualization of images

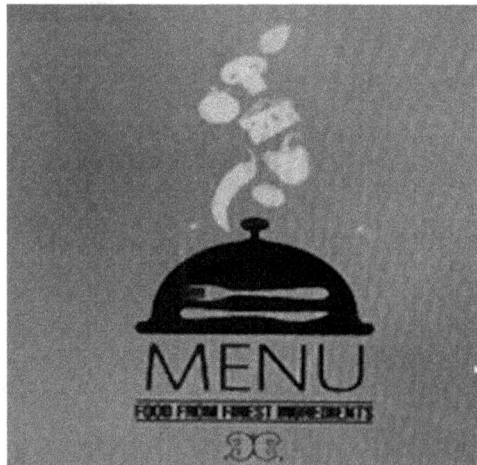

FIGURE 5.2　AR-enabled 3D visualization of menu card items (target image)

consolidating the QR code with the conventional AR framework. In this chapter, we join these two strategies to build up an item demo framework. A QR code is embedded on the menu card and afterward, a 3D virtual article created utilizing Vuforia is shown on the QR code through Unity. This framework enables the client to picture the item in a more straightforward way utilizing the camera.

5.5 IMPLEMENTATION RESULT

The implemented results of the proposed system are presented in Figures 5.2, 5.3 and 5.4.

FIGURE 5.3   AR-enabled 3D visualization of menu card items – AR view of food menu

FIGURE 5.4   AR-enabled 3D visualization of menu card item – AR view of selected food item

5.6 CONCLUSION

Augmented Reality is a breakthrough technology that could considerably ease the execution of complex operations. Augmented Reality mixes virtual and actual reality, making available to the user new tools to ensure efficiency in the transfer of knowledge for several processes and in several environments. The research community, particularly in maintenance operations, has proposed various solutions based on Augmented Reality. Augmented Reality tools have offered new perspectives and have promised dramatic improvements. On the other side, Augmented Reality is an extremely demanding technology, and at the present day, it is still affected by serious flaws that undermine its implementations in the industrial context. The benefits of AR restaurant apps are quite remarkable. In this technology era, it has become essential for every business to adopt the latest technology trends. AR is one of the most flourishing technologies that restaurants are integrating to expand and stand out from the competition.

REFERENCES

1. R. Silva, J. C. Oliveira, G. A. Giraldi. Introduction to Augmented Reality. In *National Laboratory for Scientific Computation, Av. Getulio Vargas, 333*, Quitandinha – Petropolis, January 2003.
2. Jeff K. T. Tang, Tin-Yung Au Duong, Yui-Wang Ng, and Hoi-Kit Luk. Learning to Create 3D Models via an Augmented Reality Smartphone Interface. In *IEEE International Conference on Teaching, Assessment, and Learning for Engineering (TALE)*, pp. 236–241, December 2015.
3. Mehdi Mekni, Andre Lemieux. Augmented Reality: Applications, Challenges, and Future Trends. *Applied Computational Science*, 20, pp. 205–214, 2014.
4. Mauricio Hincapié, Andrea Caponio, Horacio Rios, and Eduardo González Mendívil. An introduction to augmented reality with applications in aeronautical maintenance. In *Transparent Optical Networks (ICTON), 2011 13th International Conference on*, pp. 1–4. IEEE, 2011.
5. Huaishu Peng, Jimmy Briggs, Cheng Yao Wang, Kevin Guo, Joseph Kider, Stefanie Mueller, Patrick Baudisch, Francois Guimbretiere. *RoMA: Interactive Fabrication with Augmented Reality and a Robotic 3D Printer*. In *CHI Conference, Association for Computing Machinery*, pp. 1–12, April 2018.
6. Jishuo Yang. *Literature Survey on Combining Digital Tables and Augmented Reality for Interacting with a Model of the Human Body*, University of Calgary, 2019.
7. Apoorva Biseria, Anunay Rao. Human Computer Interface-Augmented Reality. *International Journal of Engineering Science and Computing*, 6(8), pp. 2594–2595, August 2016.
8. Nor Farhah Saidin, Noor Dayana Abd Halim, Noraffandy Yahaya. A Review of Research on Augmented Reality in Education: Advantages and Applications. *International Education Studies*, 8(13), pp.1913–9039, 2015.

6 Zeta – A Web-Based Restaurant-Suggesting Chatbot

Selvaraj Rajini

CONTENTS

6.1 INTRODUCTION

In this technologically advanced world, having human presence for monitoring all internet traffic on a site is not always feasible. Sometimes, humans fail to respond to a conversation quickly, but a chatbot never does. With busy people, their tendency is to finish work easily without consuming much time. In the evergreen field of technology, chatbots plays a major role.

Chatbots [1, 2, 3] have a wide variety of possible use cases and areas of interest for both industry and researchers. More and more companies are considering adding bots to their services. Chatbots are useful as they can answer questions and assist customers without delay and at all hours. Users can either type questions in the application or they can interact with the bot through voice commands. This is where natural language processing comes to play a major role. Natural language conversation is one of the most challenging Artificial Intelligence problems, which involves language understanding, reasoning, and the utilization of common-sense knowledge.

Several types of research have been carried out to implement chatbots in different domains. This paper [4] focuses on the working of a chatbot on various user inputs and on the internal working functionalities of a chatbot. This paper [5] tried to develop an end-to-end trainable conversational system in a task-oriented environment to suggest restaurants only in a specified range. Similarly, a survey has been

DOI: 10.1201/9781003307716-6

carried out in the paper [6] inferring about the speech conversation between humans and the chatbot and also about the techniques used during speech conversation.

The main purpose and idea of chatbots are that the computer is performing a natural language conversation [7] with human clients which should be as human-like as possible. Based on the task bot was made for, the conversations [8, 9] then usually serve some specific purpose such as searching the web, setting up appointments, etc.

Our research in this field made us collect a lot of information on how a chatbot actually works [10, 11, 12, 13] and the algorithms used to make it more effective. Naïve Bayes classification is used to categorize the words and put them in a specific class. After all the processing of data had taken place, the bot trains itself to be more effective by Machine Learning methodologies [14, 15, 16, 17, 18].

Likewise, the project on a chatbot for restaurants will make the user consume less time and help them choose the restaurant wisely based on several reviews. It accesses the location of the user and lists the restaurants nearby to them. It also displays the distance between the user's location and the restaurant's location. Through this application, the user can order their food. The bill details for the ordered food will be sent via email to the user.

The whole application provides end-to-end user access, starting from the Chat UI to sending the bill details, it has been built with great flow. Similarly, all the restaurant-suggesting chatbots are built as mobile applications and on the web, it has very limited scope and development. The idea is to develop a similar web application with all the features the same as in the mobile application.

6.2   OBJECTIVES

The main objectives are to:

- Train a bot in Dialogflow with sample utterances with respective responses
- Implement all the features to place an order in a restaurant
- Develop an end-to-end conversational chatbot to suggest restaurants

Figure 6.1 describes the workflow of the complete system. Chatbots works like a request–response cycle. A tool, Dialogflow, has been used to train the bot. In this tool, an agent should be created. An agent performs all the actions for which the bot is trained. For each agent, intents have to be created. Groups of similar actions combine together to form an intent. An intent is trained with several requests and responses, so that for each user request a respective response will be generated. Once the user sends a request, the bot responds with the appropriate message. When the user gets the response which is invoked from the nearby restaurant, the location of the user is fetched and all the restaurants will be displayed. When the user selects each restaurant, the menu of that particular restaurant will be displayed. The user can view the menu and place orders. As soon as the order gets placed, the bill details will be sent as an email to the user. This is how the whole restaurant bot – Zeta will work.

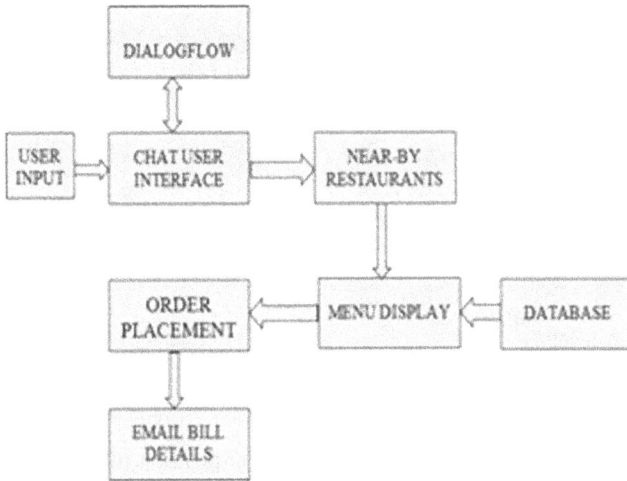

FIGURE 6.1   System architecture

6.3   MODULES

The system consists of three main modules: the chat interface module, the displaying nearby restaurants module and the displaying menu and order placement module.

6.3.1   THE CHAT INTERFACE MODULE

The user interface of the chat interface module is implemented by AngularJS. All the basic details that are required for the chat interface are implemented. Details such as message module, chat module to display the messages, etc. The chat interface module is connected using Dialogflow. The Dialogflow tool provides two tokens: the client access token and the developer access token for each agent. Chat UI is connected with Dialogflow by inserting the client access token in the angular project. The token is inserted in the project because the responses for each user request will be given by the bot i.e., the agent which is initially trained (Figure 6.2).

In Dialogflow, several intents are created for each action. Each intent is trained with a number of requests and responses based on the action which is going to be performed by the bot. Whenever the chatbot recognizes the user request, Dialogflow will map that user request to the respective intent and any random response from that intent will be shown to the user. When the user's request is related to the restaurant or food, Dialogflow will map to the nearby-restaurant intent. The nearby restaurant is trained with multiple requests and a single response i.e., getting your location. When the bot gets request from user, then it displays all the restaurants returned from the Google Maps API to the user.

6.3.2   RESTAURANT DISPLAY

Using the Google Maps API, restaurants, hospitals, ATMs, petrol stations, and several other basic and most visited places by the public can be separated. For displaying

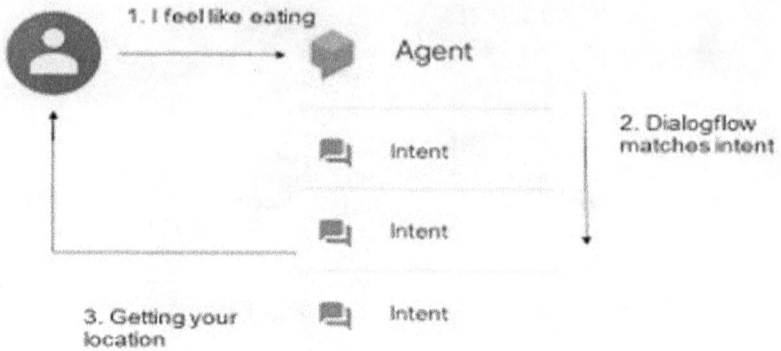

FIGURE 6.2   Agent response

restaurants, certain parameters such as the current position of the user i.e., the latitude and longitude of the user's location, the type search should be restaurants as well as the radius and the API key needs to be passed.

Radius contains the distance that the restaurants should be fetched and displayed. The API key is like a token which each user should create for the particular API used using the Google developer account.

With all these parameters as input, the Google Maps API will return a list of restaurants in json format, which then will be parsed to access each field of the returned list.

6.3.3   Menu Card Module

When the user clicks on a particular restaurant, it will be redirected to the respective restaurant's menu page. On this page, it will display the food details such as images, price, and name. The details of food and images will be stored in a MySQL database. They can be retrieved using php and stored as an array of angularJS objects. Using the ng-repeat function of AngularJS the array of objects will be looped through. An array will be initialized using ng-init function of angularJS which will call a function LoadProduct. The function will call a php page which will retrieve product data from a database using MySQL query. Connection to the database will be made using Apache server Xampp and connection strings having information about PhpMyAdmin credentials and database name. When the details are loaded they can be added to the cart by clicking add button. There will be a table displaying the food name, food quantity, and food amount. The quantity and value amount will be updated when items are added or removed. When the Add button is clicked the ng-click will call a function AddToCart which will make an ajax request and pass the object of the items array as a json object which will be parsed in the AddCart php page and will be stored as a separate object. All the product details such as name, price, and quantity will be stored in separate variables.

The cart details will be stored as a session. The session will be looped and it will be checked as a key-value pair and if the key is equal to the product id of the product added then the product quantity of the session will be increased. Otherwise, all the variables will be added to an array named $item_array and this array will be added to the session and a new row will be added to an html table. When the same product is added more than once the quantity will be increased and the product price will be calculated by multiplying quantity and price.

6.4 CONCLUSION AND FUTURE WORK

The bot is first trained with a small knowledge base which will grow dynamically as the user interacts with the bot. The bot uses Machine Learning techniques to predict the answers for the user. There will be a need for a large amount of knowledge base for the bot to answer any question asked by the user in any field. When the user asks a query, this query will be matched against the various patterns present in the knowledge base and the intent actions corresponding to that slot will be returned in form of text to the user.

Thus, the chatbot will be greatly helpful in reducing the manual task involved in finding restaurants of their preference by suggesting restaurants and keeping their personal tastes and likes in consideration. It is used for quick interactions. This will be available 24 hours to answer the users which is an advantage that allows the user to search for restaurants anywhere at any time. This will also provide an ample amount of options for restaurants from which the users can pick from. It is more user interactive as it responds to the query entered by the user very precisely because it is a domain-specific chatbot system. It also helps the user to order the needed food in advance.

The scope of the chatbot can be improved by increasing its efficiency and its performance by training the bot with several occurrences of the conversation and predicting the correct response.

6.5 RESULTS

The chatbot from Dialogflow can be integrated to make an application and used by users. It can incorporate users and menus of restaurants. It can display a menu of restaurants and these menu items can be added to the cart. The bot can track location using GPS and find restaurants within a particular distance.

REFERENCES

1. Jia, J. (2003). The study of the application of a keywords-based chatbot system on the teaching of foreign languages. arXiv Preprint cs/0310018.
2. Dhillon, R., Bhagat, S., Carvey, H., & Shribeg, E. (2004). Meeting recorder project: Dialog act labeling guide (No. ICSI-TR-04-002). International Computer Science Inst Berkeley, CA.
3. Anderson, T. (2006). Interaction in learning and teaching on the educational semantic web. In *Interactions in Online Education: Implications for Theory and Practice* (pp. 141–155), Vol. 44, No. 5.
4. Sansonnet, J. P., Leray, D., & Martin, J. C. (2006, August). Architecture of a framework for generic assisting conversational agents. In *International Workshop on Intelligent Virtual Agents* (pp. 145–156). Springer, Berlin, Heidelberg.

5. Mohamed, A. R., & Hinton, G. (2010, March). Phone recognition using restricted Boltzmann machines. In *2010 IEEE International Conference on Acoustics, Speech and Signal Processing* (pp. 4354–4357).
6. Weinberger, M. (2017). *Why Amazon's Echo Is Totally Dominating–And What Google, Microsoft, and Apple Have to Do to Catch up.* Business Insider Inc, India.
7. Sansonnet, J. P., Leray, D., & Martin, J. C. (2006, August). Tool for conversation. In *International Workshop on Intelligent Virtual Agents* (pp. 145–156). Springer, Berlin, Heidelberg.
8. Du Preez, S. J., Lall, M., & Sinha, S. (2009, May). An intelligent web-based voice chat bot. In *IEEE EUROCON 2009*, St. Petersburg, Russia, 2009, (pp. 386–391). doi: 10.1109/EURCON.2009.5167660.
9. Hung, V., Elvir, M., Gonzalez, A., & DeMara, R. (2009, October). End to end trainable ChatBot. In *2009 IEEE International Conference on Systems, Man and Cybernetics, San Antonio, TX, USA, 2009,*(pp. 1236–1241).
10. Gruhn, R. E., Minker, W., & Nakamura, S. (2011). *Statistical Pronunciation Modeling for Non-Native Speech Processing.* Springer Science & Business Media, Berlin.
11. Kumar, M. N., Chandar, P. L., Prasad, A. V., & Sumangali, K. (2016, December). Android based educational Chatbot for visually impaired people. In *2016 IEEE International Conference on Computational Intelligence and Computing Research (ICCIC)*, Chennai, India (pp. 1–4).
12. Bradesko, L., & Mladenic, D. (2012). A survey of chatbot systems through a Loebner prize competition. In *Proceedings of Slovenian Language Technologies Society Eighth Conference of Language Technologies, Ljubljana.*
13. Abdul-Kader, S. A. (2016). Survey on Chatbot design techniques in speech conversation systems. *International Journal of Advanced Computer Science and Applications(IJACSA)*, 6 (7).
14. Bhalotia, S., & Bisten, S. (2018). Implementation of ChatBot using AI and NLP. *International Journal of Innovative Research in Computer Science and Technology (IJIRCST)*, 6 (3). ISSN: 2345-5552.
15. Ciechanowski, L., Przegalinska, A., Magnuski, M., & Gloor, P. (2018). Mobile conversational commerce: Messenger chatbots as the next interface between businesses and consumers. *Future Generation Computer Systems*, 92, 539–548.
16. Sharma, V., Goyal, M., & Malik, D. (2017). An intelligent behaviour shown by chatbot system. In *International Journal of New Technology and Research (IJNTR)*, 3 (4).
17. Eeuwen, M. V. (2017). Mobile conversational commerce: messenger chatbots as the next interface between businesses and consumers, University of Twente, The Netherlands.
18. Shang, L., Lu, Z., & Li, H. (2015). *Neural Responding Machine for Short-Text Conversation.* Huawei Technologies Co. Ltd, Beijing, China.

7 Obstacle Avoidance and Image Recognition System

L. Latha

CONTENTS

7.1   INTRODUCTION

Due to the rapid development of various sensors including radars, lidars, camera systems and also wireless communication, driver assistance systems and autonomous vehicles have made significant advances in recent years [1]. The main requirements that are imposed on autonomous vehicles are the ability to cover long distances in a

DOI: 10.1201/9781003307716-7

safer way, while decreasing the rate of accidents and traffic jams, and obeying the traffic rules, all without human interaction [2]. This chapter focuses on the prototype of a driverless car. This has not yet been implemented in India. This idea has broad scope for the future. Many automotive companies are working on different parts to make it complete automated vehicle but it has not yet come into force and has not yet been integrated into a complete vehicle. This project includes four modules which help the car to drive in an automated manner. The driverless car has not been trusted by people when it was implemented in the US. Hence implementing it step-by-step by adding automotive features that increases the hope of customers.

7.2   OBJECTIVE

The driverless car has many advantages for the public. The accidents can be reduced as many accidents are due to the carelessness of the driver. In an emergency situation, it is not necessary for a person that knows how to drive be there. This project has some modules integrated to become an automated car. There is no need for a driving license. In case of emergencies, it plays a major role. The self-driving vehicle will greatly reduce traffic and parking costs, accidents and pollution emissions, and chauffeur non-drivers, reducing roadway costs and eliminating the need for conventional public transit services.

7.3   RELATED WORKS

Modern automobile companies are coming up with new autonomous features in their more recent systems. Technological advancements in every day areas such as information technology, communication data analysis, etc. are vast. The realm of autonomous cars is also progressing at a rapid rate. Various semi-autonomous features introduced in modern cars such as lane keeping, automatic braking, and adaptive cruise control are based on such systems. Extensive network-guided systems in conjunction with vision-guided features are the future of autonomous vehicles. It is predicted that most companies will launch fully autonomous vehicles by the advent of the next decade [3]. The future of autonomous vehicles is an ambitious era of safe and comfortable transportation. Segway Incorporated and General Motors both jointly developed a two-seat electric car, which could be driven normally or operated autonomously. It is known as GM's EN-V (General Motor's Electric Networked Vehicle).

The first cars licensed for autonomous driving on the streets and highways of the German city of Berlin have been made. It was developed by the Auto NOMOs Labs [4]. It was a project of Freie University, Berlin, and funded by the German Federal Ministry of Education and Research. It has a very accurate GPS unit and three laser scanners at the front, which detect any pedestrian near or around the car. It can also detect traffic lights, intercity traffic and roundabouts [5]. Karlsruhe Institute of Technology/FZI (Forschung zentrum Informatik) and Daimler made a Mercedes-Benz S-class vehicle which drove completely autonomously for 100 km. The vehicle followed the historic Bertha Benz memorial route [6]. It used next-generation radars and stereo cameras which aided it in its autonomous drive. The reduction of accidents was its main aim, caused mainly by human error. An algorithm to link various aspects of automation and machine vision was used. Toyota developed its autonomous car basically for the elimination of crashes which is one of the main causes of deaths caused by automation mishaps [3]. Toyota used something it calls ITS (Intelligent Transport Systems) technology. The car is semi-autonomous, the driver can gain control of the car anytime he/she wishes.

7.4   METHODOLOGY

There are four modules in this proposed work and the prototype for the same is shown in Figure 7.1.

7.4.1   OBSTACLE DETECTION

This module of the car will detect obstacles and drive accordingly. As an example any turns and cuts it will do without colliding, and when about to collide with stationary things it will avoid collision by turning to the other side [7]. The ultrasonic sensor is used to detect the obstacle and sends information to Arduino UNO where the action to be taken is written as code. The minimal distance requirement is given in the code to the ultrasonic sensor, so that it detects the object and avoids collision.

Obstacle detection based on visual image information is used in automated driverless cars. Obstacle detection using an ultrasonic sensor detects the obstacles by the relationship between the phase delay and the peak value of the echo and then combined it with the echo time, which had a higher detection accuracy. An example of detecting obstacles is shown in Figure 7.2.

Ultrasonic sensors work by sending out a sound wave at a frequency above the range of human hearing [8]. The transducer of the sensor acts as a microphone to receive and send the ultrasonic sound. The working principle of this module is

FIGURE 7.1   Prototype of the system

FIGURE 7.2 Obstacle detection

simple. It sends an ultrasonic pulse out at 40kHz which travels through the air and if there is an obstacle or object, it will bounce back to the sensor. By calculating the travel time and the speed of sound, the distance can be calculated.

7.4.2 COLLISION AVOIDANCE

This module of the car will detect the vehicles and stop before getting within 10m and avoid a collision. The vehicle will keep a distance of 5 to 10m from another vehicle. When the vehicle is moving it avoids colliding with a vehicle in front when it is about to stop [9]. The ultrasonic sensor is used to detect the obstacle and sends information to Arduino UNO where the action to be taken is written as code.

Collision avoidance in an automated driverless car is done using elastic band theory. The path of the vehicle is changed when the path is disturbed by external forces. Since it is not possible to avoid collision with emergency braking due to inadequate time, emergency braking is the last choice for the algorithm. When any obstacle is detected, the algorithm helps to find an alternate path to avoid collision [10]. If there is no option for an alternate path then the velocity of the vehicle is reduced to zero.

7.4.3 TRAFFIC SYMBOL RECOGNITION

This module of the car will recognize traffic signals and act accordingly. This also recognizes the traffic symbols on the road. A Raspberry Pi is used for symbol recognition and it recognizes the symbols and sends info to Arduino and it acts accordingly. Image processing, which is an artificial intelligence concept, is used. Image recognition plays a major role in it. The image processing is done by a Raspberry Pi camera which acts as a sensor and the Arduino which gives the action acts as an actuator.

The traffic symbols are recognized using the computer vision Haar-cascade algorithm as shown in Figure 7.3. The sample images are stored in the databases. The

FIGURE 7.3   Traffic symbol identification

classification algorithm is used to classify the symbols [11]. The images are trained using a train set and test set of data. The images are normalized by using test sets and histogram equalization is done.

7.4.3.1   Haar-Cascade Algorithm

The proposed system consists of the following two main stages: detection and recognition. The complete set of road signs used in our training data and recognized by the system. The system uses a Raspberry Pi as the processing engine and OpenCV as the software engine. The detection stage uses the Haar cascades based on the Haar features of an object for the detection of a traffic sign. It is a Machine Learning–based approach where a cascade function is trained from a lot of positive and negative images. It is then used to detect objects in other images. Initially, the algorithm needs a lot of positive images (images of a sign) and negative images (images without a sign) to train the classifier. Then we need to extract features from it [12]. SVM is a supervised learning method that constructs a hyper-plane to separate data into classes. The "support vectors" are data points that define the maximum margin of the hyper-plane. Although SVM is primarily a binary classifier, multiclass classification can be achieved by training many one-against-one binary SVMs. SVM classification is fast, highly accurate, and less prone to overfitting compared to many other classification methods. It is also possible to very quickly train an SVM classifier, which significantly helps our proposed method, given our large amount of training data and a high number of classes.

FIGURE 7.4 Traffic light recognition

7.4.4 Traffic Light Recognition

This module of the car will recognize traffic lights and act accordingly. This also recognizes the traffic lights on the road. A Raspberry Pi is used for color recognition and it recognizes the color and sends info to Arduino and acts accordingly. Image recognition plays a major role in it. The image processing is done by a Raspberry Pi camera which acts as a sensor and the Arduino which gives the action acts as an actuator.

The image processing algorithm is used to recognize the traffic lights. The detection step is achieved in grayscale with spotlight detection [13]. The image processing algorithm gives the best accuracy when compared with the standard object recognition method. The light is sensed with the help of a Raspberry Pi camera. The threshold of the color is used to differentiate from the different colors (Figure 7.4).

7.5 HARDWARE REQUIREMENTS

7.5.1 Raspberry Pi

Raspberry Pi uses a Linux kernel operating system. The OS for the Raspberry Pi needs to be installed and booted manually, The Raspbian Buster lite OS is installed through a memory card. The OS can be connected to an external mouse and keyboard. It behaves like a little computer.

7.5.2 Raspberry Pi Camera

The camera consists of a small circuit board which connects to the Raspberry P's camera serial interface bus connector via a flexible strap cable. We should enable camera options in the Raspberry Pi configuration settings. The camera is supported in the latest version of the Raspbian OS. It captures images and videos and converts them into YUV format images. They are command-line programs. The shell command runs the video capture programs.

7.5.3 ULTRASONIC SENSOR

The ultrasonic sensor measures distance by using ultrasonic waves. The sensor emits an ultrasonic wave and receives the wave reflected back from the target. The ultrasonic sensor has a single element for both reception and emission.

Distance=1/2×T×C

where T is the time between reception and emission and C is the speed.

7.5.4 ARDUINO UNO

Arduino UNO is an open source microcontroller board. The board has 14 digital pins, 6 analog pins and is programmable with Arduino IDE via a type-B USB cable. It can be powered by a 9 volt battery and accepts voltage between 7 and 20 volts.

7.6 IMPLEMENTATION POLICIES

7.6.1 VEHICLE PERFORMANCE GUIDANCE FOR AUTOMATED VEHICLES

"Safety Assessment" for the safe design, development, testing, and deployment of automated vehicles [14].

7.6.2 MODEL STATE POLICY

This section presents a clear distinction between Federal and State responsibilities for regulation of HAVs, and suggests recommended policy areas for states to consider with the goal of generating a consistent national framework for the testing and deployment of highly automated vehicles.

7.6.3 CURRENT REGULATORY TOOLS

This discussion outlines DOT's current regulatory tools that can be used to accelerate the safe development of HAVs, such as interpreting current rules to allow for greater flexibility in design and providing limited exemptions to allow for testing of non-traditional vehicle designs in a more-timely fashion.

7.6.4 MODERN REGULATORY TOOLS

This discussion identifies potential new regulatory tools and statutory authorities that may aid the safe and efficient deployment of new lifesaving technologies.

7.7 IMPLEMENTATION ISSUES

1) Creating (and maintaining) maps for self-driving cars is difficult work
2) Driving requires many complex social interactions – which are still tough for robots

3) Bad weather makes everything trickier
4) Cybersecurity will likely be an issue – though a surmountable one
5) Usage of different types of gestures for the same traffic sign may confuse the vehicle in some cases

7.8   CONCLUSION

This chapter discusses the basic chronology leading to the development of autonomous cars. Autonomous vehicles developed from basic robotic cars to much more efficient and practical vision-guided vehicles [15]. The driverless car becomes the next important innovation in transportation technology. Developments in autonomous cars are continuing and the software in the car is also continuing to be updated. Though it all started from the basic sensors in a car, still more semi-autonomous features will come up in the future, thus reducing the congestion, and increasing safety with faster reactions and fewer errors.

7.9   FUTURE SCOPE

The successful implementation of the prototype will create hope in the customers' minds in regard to driverless cars. Hence an increase in sales of driverless cars can be made. Favorable algorithms can be used.

REFERENCES

1. Rosique, F.; P.J. Navarro; C. Fernández, and A. Padilla. "A systematic review of perception system and simulators for autonomous vehicles research." *Sensors* 19(3) (2019): 648.
2. Anderson, James M., et al. *Autonomous Vehicle Technology: A Guide for Policymakers.* Rand Corporation, 2014.
3. Bimbraw, Keshav. "Autonomous cars: Past, present and future a review of the developments in the last century, the present scenario and the expected future of autonomous vehicle technology." *2015 12th International Conference on Informatics in Control, Automation and Robotics (ICINCO).* Vol. 1. IEEE, 2015.
4. Berlin Team. "Spirit of Berlin: An autonomous car for the DARPA urban challenge hardware and software architecture." (2007): 2010. Retrieved Jan 5.
5. Keshav Bimbraw, Mechanical Engineering Department, Thapar University, P.O. Box 32, "Patiala, Punjab, India, autonomous cars: Past, present and future. A review of the developments in the last century, the present scenario and the expected future of autonomous vehicle technology."
6. Dickmann, Juergen, et al. "Making bertha see even more: Radar contribution." *IEEE Access* 3 (2015): 1233–1247.
7. Vairavan, R., et al. "Obstacle avoidance robotic vehicle using ultrasonic sensor, Arduino controller." *International Research Journal of Engineering and Technology (IRJET)*, 2 (2018: 2140–2143,.
8. Carullo, Alessio, and Marco Parvis. "An ultrasonic sensor for distance measurement in automotive applications." *IEEE Sensors Journal* 1(2) (2001): 143.
9. Yılmaz, Esra, and Sibel T. Özyer "Remote and autonomous controlled robotic car based on arduino with real time obstacle detection and avoidance." *Universal Journal of Engineering Science* 7(1) (2019): 1–7.

10. Hosny, Ahmed, et al. "Demonstration of forward collision avoidance algorithm based on V2V communication." *2019 8th International Conference on Modern Circuits and Systems Technologies (MOCAST).* IEEE, 2019.
11. Bui-Minh, Thanh, et al. "A robust algorithm for detection and classification of traffic signs in video data." *2012 International Conference on Control, Automation and Information Sciences (ICCAIS).* IEEE, 2012.
12. Gawande, Prachi. Traffic Sign Detection and Recognition Using Open CV. *International Research Journal of Engineering and Technology (IRJET)* 4 2017.
13. De Charette, Raoul, and Fawzi Nashashibi. "Traffic light recognition using image processing compared to learning processes." *2009 IEEE/RSJ International Conference on Intelligent Robots and Systems.* IEEE, 2009.
14. Fraade-Blanar, Laura, and Nidhi Kalra. "Autonomous vehicles and federal safety standards: An exemption to the rule?." *Rand* (2017): 1–14.
15. Hadiya, Mayur Rukhadbhai. "A review paper on development of autonomous vehicle." *International Research Journal of Engineering and Technology (IRJET)* 6 2019: 445–449.

8 Town Bus Tracker

N. Suganthi

CONTENTS

8.1 INTRODUCTION

Smartphones have become an integral part of our day-to-day life and smartphones are nothing without applications. These days, Android has become very popular in the market. It is useful, user friendly, and saves a lot of time and physical effort. This chapter gives an approach to one such application using Android. The bus schedule tracking application is an Android application for smartphones. It is GPS technology equipped. One must enter the pickup location and the destination location, and the application will list down the possible buses that can be taken to reach the required destination. There is also a provision to set reminders. Once the reminder is set the application will notify the user about the same, minutes prior to the departure of the bus. This application also runs in the background so one can continue to do whatever they were up to. In case of emergency, there is an alert button, which when clicked will be directed to a police station.

This is an Android-based application which is simple to use and thus it can be used by everyone with ease. Travel enthusiasts will find this application very useful. Even people who use town buses on a regular basis will find this it handy. This app also helps to save a lot of time and avoid confusion. Apart from frequent and infrequent travelers, this would also help tourists to experience a comfortable vacation.

DOI: 10.1201/9781003307716-8

8.2 LITERATURE SURVEY

A Bus Tracking System (BTS) is proposed in paper [1] and mentions an increase in rural and urban populations, fuel rates, and pollution as leading the way for the public transportation system. Town buses are mostly used by local people, especially by the working community. Hence there is a need for a BTS to make their travels smooth and easy by reducing the waiting times and unwanted confusion for the passengers.

A GPS-based BTS, enables GPS on the bus to gather information minute-by-minute according to the needs of the user. Integrating the Google API with the application will enable the passengers to track the exact location of the bus. Instead of waiting at the bus stop unnecessarily this GPS-enabled BTS will help to save time and to use the easiest and clearest route and provides timely bus service.

8.3 PROPOSED WORK

8.3.1 SOFTWARE MODULE

8.3.1.1 Android Application for Town Bus Scheduler

This project is a user-friendly interface. As mentioned in paper [2] the Android-based smartphone application displays information about buses, all of the possible routes for the bus, tracks the current location of the bus and display the fixed route, estimates the bus arrival time and frequency of buses. It can also alert about chosen buses via SMS as mentioned in paper [3] and provide security systems for passengers. This project is to overcome all the drawbacks faced in already existing applications (Figure 8.1).

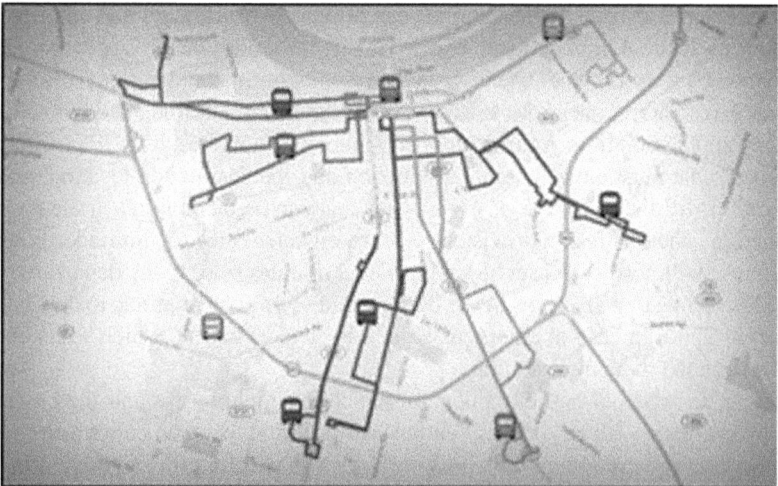

FIGURE 8.1   Bus route design

8.3.1.2   Safety and Alert System

A town bus scheduler can manage to provide both a safety and alerting system to the users. Many passengers feel exposed to harassment and public disturbances while traveling on the public transport system. To overcome such problems our application helps to alert nearby police stations and family members about current bus details as well as the current location.

8.3.1.3   Tracking System

Installing a GPS tracking system not only improves the efficiency of routes but also shows the live status of the buses. As shown in paper [4] this application provides both fixed routes and live tracking of routes. Mainly this system allows you to plan routes and helps to avoid delays.

The flowchart in Figure 8.2 provides an understanding of the Town Bus Scheduler (TBS) application. In the "Bus" selection tab after selecting the required bus, users can see the bus information. This system provides tracking information about the bus and displays maps to see nearby stops and bus routes. Passengers from new cities and tourists from other places can easily get their current positions and bus routes by using this Android application (Figure 8.3).

8.3.2   Hardware Module

8.3.2.1   Implementing Global Positioning System (GPS)

Smartphone users using GPS can receive navigation instructions that tell them where they are and they can get the exact location of the transit. As shown in paper [5] GPS

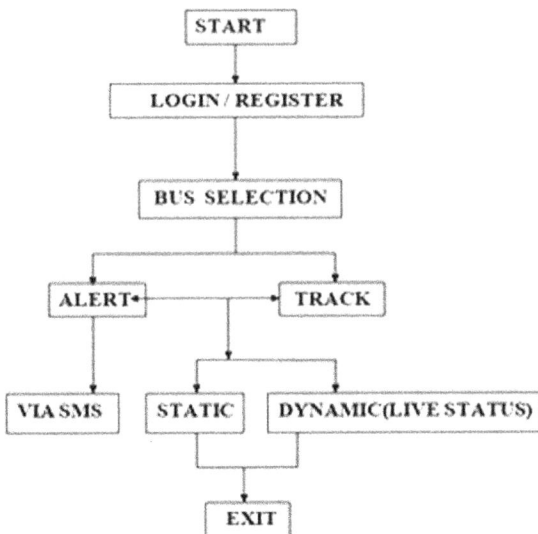

FIGURE 8.2   Flow chart

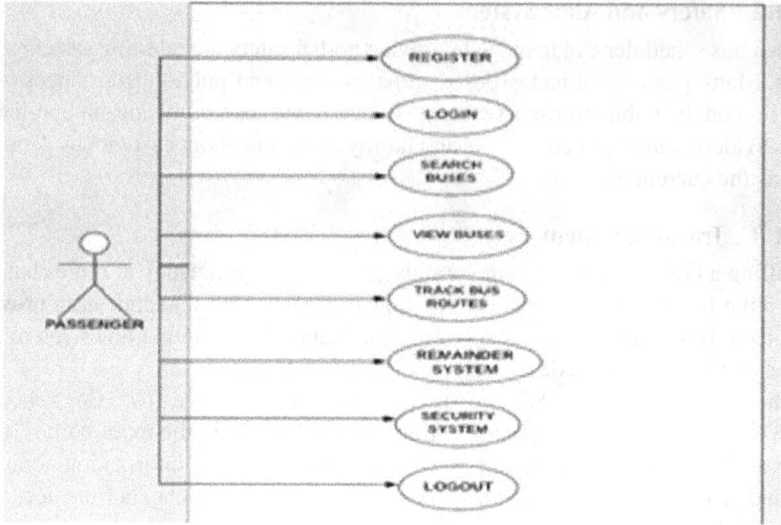

FIGURE 8.3 UML diagram

sends information to a database within the GPS unit. Working on a GPS needs a modem for communication with the smartphone.

8.3.2.2 Arduino Functions in TBS

Arduino is an open-source platform for both hardware and software services. The main purpose of Arduino is to control the whole process with GPS service and GSM module, GPS receiver for tracking the points of the vehicle and GSM module for sending the points to users by SMS.

8.3.2.3 Implementation of Global System for Mobile Communication (GSM)

The Town Bus Scheduler (TBS) uses a GSM module to activate the communication between the GPS service and the smartphone. This system is similar to one mentioned in paper [5] to get the input details (tracking points) from Arduino received from GPS service and compares the information and sends the received information to smartphone devices.

The block diagram, Figure 8.4, shows the module design of the Town Bus Scheduler where Arduino acts as a controlling unit to transfer information from a GPS receiver to a GSM module to reach the smartphone.

8.4 CONCLUSION

With an alarming increase in air pollution in cities such as Delhi, there is an increased need to reduce pollution as soon as possible. With the failure of scheme that is implemented in Delhi, private vehicles with registration plates ending in odd

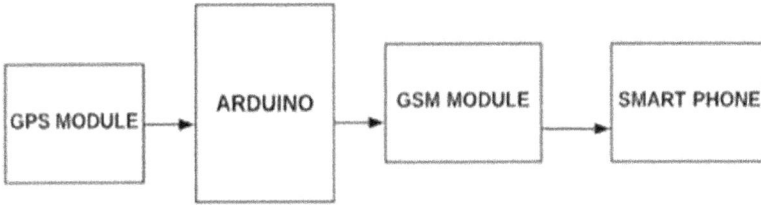

FIGURE 8.4   Block diagram

numbers could ply on odd dates, and even numbers on even dates, one of the most efficient ways to control air pollution without affecting the economy for both the government and the public is the usage of public transport on regular basis. Time is a very important aspect of everyone's life, this application will come in handy for those who are running around with busy schedules.

8.5   RESULT

An Android application with the four modules described is implemented. With the use of GPS and GSM technologies, real-time location is tracked and displayed for passengers. This application will make the public transport system more passenger-friendly.

8.6   FUTURE WORK

This application could further be developed to detect crowding in transit using cameras or sensors. This would help the public to schedule their travel right from their seat and saves waiting time at bus stops.

REFERENCES

1. Manish Chandwani, Lokeshjeswani Bhoomika Batheja, Praveen Devanani, 2018, "Real Time Bus Tracking System", *IOSR Journal of Engineering*, 14, 24–28.
2. S. Suganya, A. Valarmathi, 2017, "GPS Enabled Android Application for Local Bus Schedule System", *IJSRCSEIT*, 2(3), 48–52.
3. R. Maruthi, 2014, "SMS based Bus Tracking System using Open Source Technologies", *International Journal of Computer Applications*, 86, 44–46.
4. Marko Celana, Marjan Lep, 2017, "Bus Arrival Time Prediction Based on Network Model", *Procedia Computer Science*, 113, 138–145.
5. Prafull D. Patinge, N.R. Kolhare, 2012, "Smart Onboard Public Information System Using GPS & GSM Integration for Public Transport", *International Journal of Computers and Applications*, 1, 308–312.

9 Intelligent Predictive Analytics in Farming Using IoT in Big Data

S. Sangeetha and N. Suganthi

CONTENTS

9.1 INTRODUCTION

The continuous generation of data from multiple sources has created numerous opportunities across different domains such as intelligent transport systems, smart environments, smart health, smart agriculture, etc. The Internet of Things (IoT) also plays a major role in smart agriculture which has various embedded computing devices such as sensors. IoT connects the values obtained by the sensors and controls remotely through the Internet. Data collected and stored continues to grow exponentially and is increasingly everywhere and in many formats. Traditional solutions are failing under new requirements. Big Data analytics is an advanced analytic technique against very large, diverse data sets that include structured, semi-structured and unstructured data, from different sources, and in different sizes. Advanced analytics techniques are text analytics, Machine Learning, predictive analytics, data mining, statistics, and natural language processing. Some research challenges in IoT are that it is difficult to capture data fully the first time and also data captured is not reliable. Research challenges in Big Data are too much data to analyze effectively and also storing and retrieving large volumes of data. Now, the number of things connected to the Internet is greater than the number of people living on Earth. By 2020, the number of things connected to the Internet will be more than fifty billion items. 94% have faced challenges in collecting and analyzing IoT data. The

DOI: 10.1201/9781003307716-9

government has taken an initiative with a vision of developing a connected, secure, and smart system based on our country's needs. Smart agriculture is one such vision. To survive in the competitive and rapidly changing agriculture market of the twenty-first century, a farmer's passion for working on the land and hard work is no longer enough. Innovative farming technologies are required to increase the quantity and quality of crops. Farmers need automatic monitoring of the condition of the field without going to the field and to be able to make intelligent and profitable decisions.

9.2   LITERATURE SURVEY

Table 9.1 gives a summary of various technologies used in smart farming systems along with their inference.

9.3   PREDICTIVE ANALYSIS IN FARMING

The various challenges faced by the Internet of Things are security, connectivity, compatibility and longevity, technology standards, and intelligent analysis actions.

- Security – Many new nodes are created and act as a malicious actor trying to carry out a DoS attack. Monitoring the public through surveillance camera footage, this acts as a passive attacker.
- Connectivity – A lot of devices are connected to a server in a bottleneck. It needs to be decentralized by fog/edge computing.
- Compatibility and longevity – While connecting various devices, it requires the deployment of extra hardware and software due to a lack of standardized machine-to-machine protocols.
- Technology standards – It doesn't have standards for handling unstructured data. A lack of technical skills to leverage newer aggregation tools.
- Intelligent analysis actions – which relate to cognitive skill. It uses AI techniques, fuzzy logic, etc. to take intelligent actions. The problems faced are inaccurate legacy systems' ability to analyze unstructured data and legacy systems' ability to manage real-time data.

A huge amount of data is collected by various sensors over a period of time, which leads to Big Data systems. It has the potential to improve predictive models. A predictive model uses various Machine Learning algorithms to study data collected by various types of sensor and give predictions about the possible outcomes. This predictive system uses various platforms or tools to analyze the data collected from sensors and give a decision.

Figure 9.1 shows the workflow of predictive analysis in farming. Various sensors are deployed at different locations in the field to measure real-time data. Values generated continuously from various sensors at multiple locations over a period of time will lead to Big Data (Jiang et al., 2010). These need to be analyzed using Machine Learning techniques and compared with a knowledge base. The agricultural datasets are collected from various heterogeneous resources over several years – agricultural

TABLE 9.1
Survey of Smart Farming Using Various Technologies

Author	Inference
Manijeh Keshtgari et al. (2012)	**Title: A Wireless Sensor Network Solution for Precision Agriculture Based on ZigBee Technology** *(Research Gate)* **Technology: Zigbee** Real-time data of climatologically and other environmental properties were sensed using ZigBee wireless technology and control decisions were taken based on it.
Sriharsha et al. (2012)	**Title: Monitoring the paddy crop field using zigbee Network** *(Research Gate)* **Technology: Zigbee** It focuses mainly on sensing and monitoring the temperature, humidity, and water level of paddy crop fields and gives various sense analyzes in paddy crop fields.
Agboola et al. (2013)	**Title: Development of a Fuzzy Logic-Based Rainfall Prediction Model** *(Research Gate)* **Technology: Fuzzy Rule base** It predicts rainfall using a Fuzzy inference system. PE, RMSE, and MAE values on data were computed and found comparatively less.
Mohanraj et al. (2016)	**Title: Field Monitoring and Automation using IoT in Agriculture Domain** *(Elsevier)* **Technology: IoT** Created a knowledge dataflow model to connect various scattered sources to the crop structures. Replaced manual procedures with the IoT.
Sherine et al. (2013)	**Title: Precision farming solution in Egypt using the wireless sensor network technology** *(Elsevier)* **Technology: WSN** WSN in cultivating the potato crop in Egypt and also stated how the automation of agriculture using a wireless sensor network will help to solve a lot of Egyptian agricultural problems and improve the crops.
Hemlata Channe et al. (2015)	**Title: Multidisciplinary Model for Smart Agriculture using Internet of Things (IoT), Sensors, Cloud-Computing, Mobile-Computing & Big Data Analysis** *(Research Gate)* **Technology: IoT & BIG DATA** It increases agricultural production and controls the cost of agro-products by using various technologies such as IoT, sensors, cloud-computing, mobile-computing, and Big Data analysis.
Kuwata (2015)	**Title: Estimating crop yields with deep learning and remotely sensed data** *(IEEE)* **Technology: Deep Learning** The accuracy of crop yield has been improved by integrating more data, such as soil properties, irrigation and fertilization, into the input dataset.
Gandhi et al. (2016)	**Title: Rice crop yield prediction in India using support vector machines** *(IEEE)* **Technology: SVM** Applied an SMO classifier using the WEKA tool on the dataset to predict crop productivity in different climatic conditions.
Ruchita Thombare et al. (2017)	**Title: Crop Yield Prediction Using Big Data Analytics** *(IEEE)* **Technology: Big Data Analytics** Reduced storage by Big Data analytics-K means clustering and an apriori algorithm.

(Continued)

TABLE 9.1 (CONTINUED)

Survey of Smart Farming Using Various Technologies

Author	Inference
Majumdar et al. (2017)	**Title: Analysis of agriculture data using data mining techniques: application of Big Data** *(Springer)* **Technology: Data Mining Techniques** Suggested data mining techniques such as CLARA and DBSCAN to obtain the optimal climate requirement of wheat such as the optimal best temperature range, worst temperature, and rainfall to achieve higher production of wheat crop. DBSCAN gave a better clustering quality. It also suggested extending the work to analyze the soil and other factors for the crop and to increase crop production under different climatic conditions.
Suvidha Jambekar et al. (2018)	**Title: Application of Data Mining Techniques for Prediction of Crop Production in India** *(Science Direct)* **Technology: Fuzzy rule–based** Suggested data mining techniques – multiple linear regression, random forest regression, and support vector regression for rice, maize, and wheat.

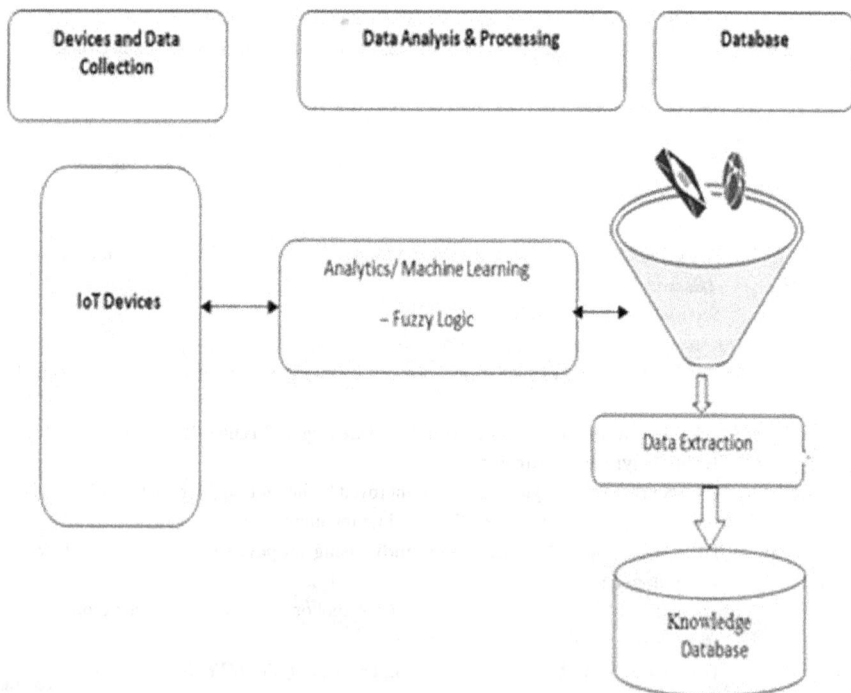

FIGURE 9.1 Flow of predictive analysis

researchers, email, web crawling, and social media. These diverse datasets may be structured, semi-structured, and unstructured from different resources. These datasets need to be analyzed using Big Data analytics, Machine Learning, predictive analysis, data mining, etc. In order to synthesize Big Data and to communicate among devices using IoT, Machine Learning techniques are employed.

In order to have accurate decisions, the following metrics need to be calculated:

   i. *Recall*
  ii. *Precision*
 iii. *F1 Score*

Based on the outcome, the values are plotted using confusion matrixes, receiver operating characteristic (ROC) curve and area under the curve (AUC) to calculate the above metrics. Here the predictive modeling system is designed for smart farming and the various steps involved are explained below.

The overall architecture of the fuzzy-based crop prediction is shown in Figure 9.2. There are two stages in fuzzy prediction. Stage 1 predicts the climate based on temperature and humidity. Temperature sensors measure the temperature of the atmosphere over time and are stored in a buffer. Similarly, the humidity sensor

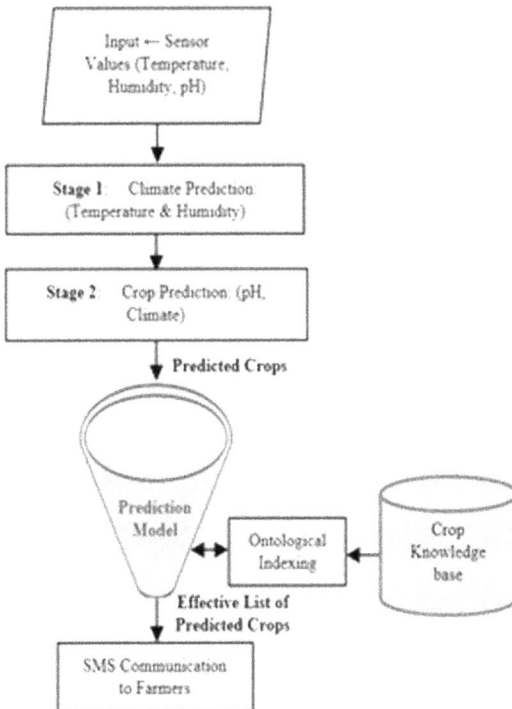

FIGURE 9.2 Workflow of crop prediction

measures the humidity of the soil stored in the buffer. Temperature and humidity values are given as input for climate prediction. Since sensors measure different values for a time interval, the measurement leads to fuzzy. Climate prediction involves various steps such as fuzzification, fuzzy rule formation using FAM and defuzzification (Kang and Gao, 2013; Gelian et al; Sakthipriya, 2014).

Stage 2 predicts the list of crops to be grown based on the predicted climate at Stage 1 and the pH value of the soil. Crop prediction involves fuzzification, fuzzy equations, and defuzzification. A list of crops predicted by Stage 2 will be tested with the crops knowledge base and predicts the effective list of crops to be grown as shown in Figure 9.3. The algorithm (Sangeetha et al., 2015 and Zimoh et al., 2013) involved in crop prediction using fuzzy logic is given as follows:

Algorithm: Crop prediction using fuzzy rules

1. Convert temperature in Celsius to %.

   - *Split the input temperature into two boundary values as min and max values.*

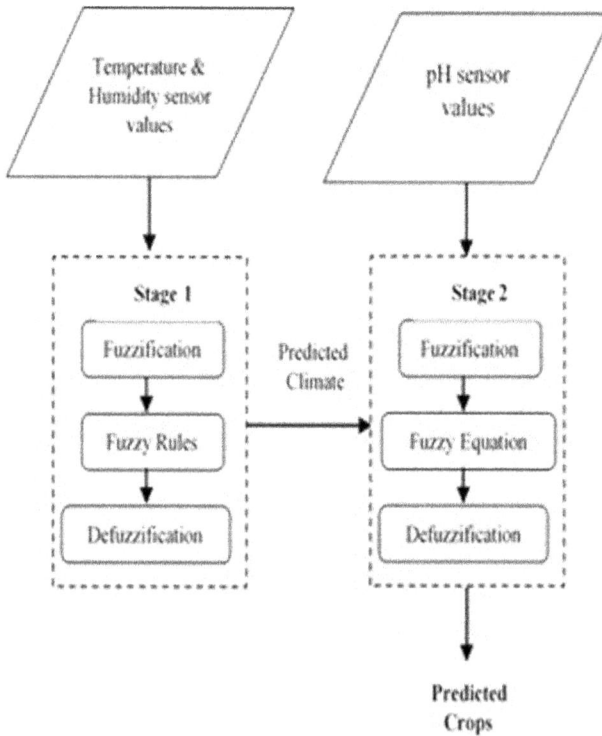

FIGURE 9.3   Fuzzy model

- *Find the average of the min and max values.*
- *Find temperature in % using the following formula:*

Temperature = original value – minimum value / average value

2. Normalize temperature and humidity values from 0 to 1 using the following formula:

Normalization of (x)=(x–min)/(max–min)

3. Convert it into linguistic variables.
4. Evaluate it using fuzzy rules to find climate.
5. Predict a list of crops to be grown using the climate and pH value of the soil.

Predicted outcomes will sometimes be confusing. In order to have accurate decisions the following performance metrics need to be followed.

9.3.1 Four Outcomes of Binary Classification

i **True positives:**
- data points labeled positive are really positive
ii **False positives:**
- data points labeled positive are really negative
iii **True negatives:**
- data points labeled negative are really negative
iv **False negatives:**
- data points labeled negative are really positive

9.3.2 Performance Metrics

i. **Accuracy**

$$Accuracy = (TP+TN) / (TP+FP+FN+TN)$$

ii. **Recall**

$$Recall = TP / (TP+FN)$$

iii. **Precision**

$$Precision = TP / (TP+FP)$$

iv. **F1 score**

$$F1\ Score = 2 \times (Recall \times Precision)\ /\ (Recall + Precision)$$

v. **Specificity**

$$Specificity = TN\ /\ (TN+FP)$$

9.3.3 Visualizing Recall and Precision

i. *Confusion matrix*: shows the actual and predicted labels from a classification problem
ii. *ROC curve*: For classifying a positive, plot the true positive rate versus the false positive rate
iii. *AUC*: calculate the overall performance of a classification model based on the area under the ROC curve

9.4 BIG DATA VALUE CHAIN

The Big Data value chain is mentioned in Figure 9.4. Each and every stage of this value chain need standardized approaches as mentioned below (Othman and Shazali, 2012.),

i. **Collection** – Structured, unstructured, and semi-structured data from multiple sources
ii. **Ingestion** – Loading vast amounts of data onto a single data store
iii. **Discovery and Cleansing** – Understanding format and content; clean up and formatting
iv. **Integration** – Linking, entity extraction, entity resolution, indexing, and data fusion
v. **Analysis** – Intelligence, statistics, predictive, and text analytics and Machine Learning
vi. **Delivery** – Querying, visualization, and real-time delivery on enterprise-class availability

9.5 MACHINE LEARNING TECHNIQUES

Machine Learning techniques serve as a basis for predictive analysis. Various Machine Learning techniques include:

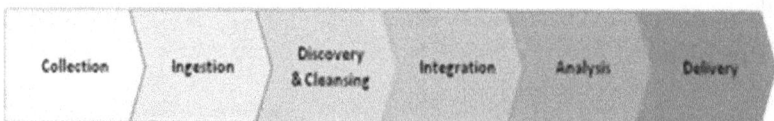

FIGURE 9.4 Big Data value chain

Regression analysis – Statistical method – examines the relationship between two or more variables. Regression models involve the following components: The unknown parameters are often denoted as a scalar or vector. The independent variables are observed in the data and are often represented as a vector. The dependent variables observed in data are often represented using the scala. The error terms, which are not directly observed in data and are denoted, use the scalar.

Clustering – Groupings of similar data.

Clustering methods
- – Density-based methods
- – Hierarchical-based methods
- – Partitioning methods
- – Grid-based methods

Bayesian methods – Dynamic analysis of a sequence of data. Unlike classical statistical methods, Bayesian statistical methods for the analysis of agricultural data directly incorporate expert agriculture knowledge in estimating unknown parameters.

Decision trees – A predictive modeling tool which has applications across a number of different areas. Decision trees are constructed via an algorithmic approach that identifies ways to divide a data set based on diverse conditions. It is the most widely used method for supervised learning. Decision trees are a non-parametric supervised learning method used for both classification and regression tasks. The goal is to create a model that predicts the value of a goal variable by learning simple decision rules inferred from the data features.

Random forest – Used to build predictive models for both classification and regression problems. The random forest consists of a large number of individual decision trees that operate as a group. Each individual tree in the random forest spits out a class prediction and the class with the most options becomes our model's prediction.

Support vector machine – A discriminative classifier formally defined by a separating hyperplane. A support vector machine is a supervised Machine Learning algorithm used for both classification and regression challenges. In this algorithm, each data item is plotted as a point in n-dimensional space with the value of each feature being the value of a particular coordinate. Then, classification is performed by finding the hyperplane that separates the two classes exactly.

Reinforcement learning – find the best possible behavior or path it should take in a specific situation. It differs from supervised learning in that labeled input/output pairs need not be presented and sub-optimal actions need not be explicitly corrected.

Deep learning – A broader family of Machine Learning methods based on artificial neural networks. Deep learning is a subdivision of Machine Learning which deals with algorithms inspired by the structure and function of the brain called artificial neural networks. It reflects the functioning of our brains. Deep learning algorithms are analogous to our nervous system's structure where each neuron is connected to each other and passes information. Deep learning models tend to perform well with the amount of data whereas old Machine Learning models stop improving after a saturation point. The main difference between Machine Learning and deep learning models is feature extraction. Feature extraction is done by humans in Machine Learning whereas deep learning models figure it out by themselves.

9.6 CONCLUSION

This proposed work is intended to increase the productivity of crops and reduce the number of workers in the field. To make profitable decisions, farmers need guidance throughout the entire farming cycle at the right time. The information required by farmers is scattered in various places which including real-time information. Using the IoT, the world is becoming more automated by replacing manual procedures, since it is energy efficient and involves minimal manpower. This project proposes automated prediction of the crop with the advantages of having ICT in the Indian agricultural sector. It shows a path for rural farmers to replace some of the conventional techniques. This proposed work overcomes the limitations of traditional agricultural procedures by utilizing water resources efficiently and also by reducing labor costs. The agricultural system is also automated to monitor regularly for watering.

REFERENCES

Agboola, A. H., A. J. Gabriel, E. O. Aliyu, B. K. Alese, "Development of a Fuzzy Logic Based Rainfall Prediction Model", *International Journal of Engineering and Technology*, 3(4), pp. 427–435, April 2013.
Channe, Hemlata, Sukhesh Kothari, Dipali Kadam, "Multidisciplinary Model for Smart Agriculture using Internet-of-Things(IoT), Sensors, Cloud-Computing, Mobile-Computing & Big-Data Analysis", *Journal ofcomputer Technology and Applications*, 6(3), pp. 374–382, 2015.
Gandhi, N., L. J. Armstrong, O. Petkar, "Rice Crop Yield Prediction in India Using Machine Learning Techniques", *International Joint Conference on Computer Science and Software Engineering (JCSSE)*, 5(1), pp. 501–507, 2019.
Jambekar, Suvidha, Shikha Nema, Zia Saquib, "Application of Data Mining Techniques for Prediction of Crop Production in India", *International Journal of Innovations & Advancement in Computer Science*, 7(4), pp. 66–69, 2018.
Jiang, X., G. Zhou, Y. Liu, Y. Wang, "Wireless Sensor Networks for Forest Environmental Monitoring", *Innovations and Trends in Environmental and Agricultural Informatics*, pp. 2–5, 2010.
Jimoh, R. G., M. Olagunju, I. O. Folorunso, M. A. Asiribo, "Modeling Rainfall Prediction Using Fuzzy Logic", *International Journal of Innovative Research in Computer and Communication Engineering*, 1(4), pp. 929–936, June 2013.
Kang, J. C., J. L. Gao, "Application of Ontology Technology in Agricultural Information Retrieval", *Advanced in Materials Research*, 756–759, pp. 1249–1253, 2013.

Keshtgari, Manijeh, Amene Deljoo, "A Wireless Sensor Network Solution for Precision Agriculture Based on ZigBee Technology", *Scientific Research Journal on Wireless Sensor Network*, 4, pp. 25–30, 2012.

K. Kuwata and R. Shibasaki, "Estimating crop yields with deep learning and remotely sensed data," *2015 IEEE International Geoscience and Remote Sensing Symposium (IGARSS)*, pp. 858–861, 2015.

Majumdar, J., S. Naraseeyappa, S. Ankalaki, "Analysis of Agriculture Data Using Data Mining Techniques: Application of Big Data", *Journal of Big Data*, 4(1), p. 20, 2017.

Mohanraj, I., Kirthika Ashokumar, J. Naren, "Field Monitoring and Automation Using IOT in Agriculture Domain", *Procedia Computer Science*, 93, pp. 931–939, 2016.

Othman, Mohd Fauzi, Khairunnisa Shazali, "Wireless Sensor Network Applications: A Study in Environment Monitoring System", *International Symposium on Robotics and Intelligent Sensors, Procedia Engineering*, 41, pp. 1204–1210, 2012.

Sakthipriya, N., "An Effective Method for Crop Monitoring Using Wireless Sensor Network", *Middle-East Journal of Scientific Research*, 20(9), pp. 1127–1132, 2014.

Sangeetha, S., M. K. Dharani, B. Gayathridevi, R. Dhivya, P. Sathya, "Prediction of Crop and Intrusions Using WSN", *Proceedings of 3rd International Conference on Advanced Computing, Networking and Informatics in the series of Smart Innovation, Systems and Technologies*, 44, pp. 109–115, 2015.

Sherine, M., Basma M. Abd El-kader, Mohammad El-Basioni, "Precision Farming Solution in Egypt Using the Wireless Sensor Network Technology", *Journal of Egyptian Informatics*, 14(3), pp. 221–233, November 2013.

Song, Gelian, Maohua Wang, Xiao Ying, Rui Yang, Binyun Zhang, "Study on Precision Agriculture Knowledge Presentation with Ontology", *AASRI Conference on Modelling, Identification and Control, AASRI Procedia*, 3, pp. 732–738, 2012.

Sriharsha, K., T. V. Janardhana Rao, A. Pravin, K. Rajasekhar, "Monitoring the Paddy Crop Field Using Zigbee Network", *International Journal of Computer and Electronics Research*, Volume 1, Issue 4, pp. 202-207, 2012.

.

10 Intelligent Transportation System
A Review of VANET Applications for Urban Areas, Technologies, and Protocols

J. Cynthia, G. Sakthipriya,
Jayaseelean Clement Sudhahar, and M. Suguna

CONTENTS

DOI: 10.1201/9781003307716-10

99

10.1   INTRODUCTION

Intelligent Transportation Systems (ITSs) provide services related to transportation to make the transportation network smarter. ITS lets the user know about the transportation scenario, which enables the creation of a more safer and integrated application of traffic networks. This can increase safety, mobility, and efficiency of the network.

ITS is one of the applications of a Vehicular Adhoc Network (VANET). VANET consists of mobile nodes and they are capable of establishing wireless communication among themselves. VANET allows communication from Vehicle-to-Vehicle (V2V), Vehicle-to-Infrastructure (V2I) and Infrastructure-to-Infrastructure (I2I). This communication is done by either an On Board Unit (OBU) or Road Side Unit (RSU). OBU is used for communication between one vehicle to another vehicle within the short transmission range. RSU is used for communication among vehicles with a long transmission range.

The transformation of a smart city mainly includes transportation. Potential problems in driving a vehicle include the analysis of traffic dynamics in the transportation network, accidents due to merging of lanes, blind curves, T-junctions and vehicle slippage during wet road conditions, traffic congestion, etc. This chapter discusses the problems by giving an overview of VANET applications. This also covers possible VANET protocols for the implementation of any kind of VANET application. Therefore, ITSs using VANET can assist driver safety.

10.2   URBANIZATION EFFECTS AND SMART CITY SOLUTIONS

Urbanization is considered to be a crisis for the following reasons. Urban cities face several challenges such as traffic congestion, poor road safety, and insufficient parking space. Traffic congestion may lead to additional fuel consumption, delay in travel time, and incurring more costs.

Smart cities offer different benefits, from smart parking to route diversion alerts. A driving assistance application provides optimized traffic signal control to reduce traffic congestion and automatic toll and fare collection to prevent delays to travel time.

Safety measures such as speed limits or disaster alerts can be sent to vehicles to avoid the loss of many lives from accidents.

Driver monitoring as well as vehicle monitoring is possible by several Internet of Things (IoT) sensors and Global Positioning System (GPS)/GPRS technologies.

So smart city technology can make the city more secure, faster, and more efficient than existing traffic.

10.3   TRAFFIC MANAGEMENT

There is a lot of literature available related to Intelligent Transportation Systems (ITSs) using VANET.

10.3.1   Traffic Management in ITS

Maslekar et al. [1] proposed a traffic control system. It is based on car-to-car communication. A direction-based clustering algorithm in VANET (DBCV) was developed. The number of vehicles reaching the intersection is gathered. Direction parameters are considered based on GPS and digital maps. Terms such as green interval, inter-green interval and red interval are considered to find traffic cycle time and optimum cycle time by modifying Webster's formula. For simulation purposes, the NCTUns simulation tool is used. Finally, a pre-timed traffic signal system was compared with a proposed traffic signal control system which indicates that the pre-timed signal had exponential waiting time behavior whereas the proposed system has a linear waiting time behavior.

Lykov and Asakura [2] proposed a tensor-based approach to detect anomalous traffic patterns. Continuum modeling is used for validation. The targeted region in the urban area is separated into mesh cells. The average speed of the vehicle inside the mesh cell at different time intervals during the day is calculated. Then it is transformed into a tensor depiction by constructing a third-order tensor. With the help of tensor robust principal component analysis, the constructed tensor is used to find anomalous traffic behavior. Continuum modeling is then implemented in urban areas by introducing probes. Probes are introduced to evaluate the performance of the proposed tensor-based approach for anomalous traffic pattern detection under different spatial resolutions and populations of probes.

Watanabe et al. [3] developed a platform called DynamicMap 2.0 which is a city-scale traffic management platform. The platform considers static information, dynamic information, road maps, and predicted information. Links, lanes, and physical shapes are the three levels of granularities considered in the road map. In order to test the platform, the prototype of cooperative merging assistance was implemented. In cooperative merging assistance, the ego vehicle is controlled by a driving simulator and the other vehicles are controlled by a traffic simulator. Embedded devices such as cameras and GPS are connected to the edge server by means of wireless communication. Edge servers and cloud servers are connected by wired communication. Dynamic map prototype was developed using the C++ language with 81 modules and it works on a Linux-based system. The developer of the traffic application writes a query which can be either a continuous query or a one-shot query. The query specifies the integration and filtering of data.

Mizuno et al. [4] proposed a distributed Constraint Satisfaction Problem (CSP). It is implemented in urban traffic signal control and the whole system was defined as a multi-agent system. Adjusting control parameters were represented as CSP. The CSP was extended to distributed CSP in which the intersection agent receives the congestion information from the road agent and changed values from the adjacent intersection agent. Based on these values, the agent reassigns values to constraints which can resolve the local congestion. Agent-oriented urban traffic simulator was developed. The developed simulator is tested on road networks which results in reduced traffic jams.

Choffnes and Bustamante [5] implemented a traffic model in VANET using Street Random Waypoint (STRAW). It was defined as a mobility model. The components of STRAW included a street placement model, a traffic control implementation which can manage the routes and their implementation.

10.3.2 Accident Avoidance in ITS

Osafune et al. [6] analyzed the accident risk from driving behavior. An application running over smartphones was developed, which works in each vehicle during driving and collects accelerometer readings such as timestamps, 3-axis accelerations, and GPS locations. More than one year of driving data (about 15 months) from more than 800 drivers were collected. The experiment was performed to explore the Accident Risk Index (ARI) which statistically separates safe drivers and risky drivers by identifying the statistically significant driving behavior which correlates with drivers' accident history. ARI was calculated by considering parameters such as sudden acceleration, sudden braking, sharp left turn, and sharp right turn.

Devi et al. [7] proposed a mechanism for providing a voice alert for avoiding accidents on merging lanes, blind curves, and T-junctions. An ultrasonic sensor is used to detect the distance between the vehicles approaching an intersection. XBee is used as a wireless communication module. The presence or absence of a vehicle is intimated to drivers by means of signaling. The red signal indicates the approach of a vehicle on the other side of the lane and the green signal indicates the non-approach of a vehicle on the other side of the lane. When the vehicle approaches closer to the intersection a pre-recorded voice alert message and buzzer are activated.

Enriquez Jr et al. [8] developed software to track the vehicles for identification and plotting of slippery road conditions which could avoid accidents. A device called CAN-BUS OBD Programmable-Expandable Network-Enabled Reader (CANOPNR) was developed. It acts as a vehicular interface and networking device. The detection was done through the vehicles equipped with CANOPNR. The software for the CANOPNR was designed using Arduino UNO IDE. The CANOPNR device reads data from the vehicle's ECU through the CAN-BUS shield and sends the data back to the server. Parameters such as steering wheel angle, speed of the vehicle, speed of each wheel, and brake pressure are obtained to detect an ABS event. An ABS event detection algorithm was developed using the above-mentioned parameters. If an ABS event is triggered with only light brake pressure, it is concluded that the area where the event took place is dangerous to drive. An ABS event with color-coded priority is displayed in Google Maps indicating the volume of slippage events.

Bhumkar et al. [9] proposed an idea to avoid accidents on highways. A real-time prototype using ARM7 was designed to control the vehicle speed and to identify the fatigue symptoms of drivers. Sensors such as eye blink, gas, alcohol, and fuel-level sensors are used. GPS and Google Maps APIs are used for software interfaces to identify a location. When the accident possibilities are detected micro controller send the SMS to driver via GSM.

10.3.3 PEDESTRIAN ACCIDENT AVOIDANCE

Sam et al. [10] described a novel idea to improve pedestrian safety. Global Positioning System (GPS) is used as a body sensor and it helps to track the location of pedestrians. The tracked location is transmitted to the nearby Road Side Unit (RSU). RSU alerts the motorist about the location of pedestrians by means of IEEE 802.11 standard communication. It is concluded that pedestrians are in danger when the vehicle speed is as low as 11km/hr when no alert was given. The proposed system indicates that pedestrians are in danger only when the vehicle speed is above 36km/hr.

10.3.4 COLLISION ACCIDENT AVOIDANCE

ElBatt et al. [11] implemented the Forward Collision Warning (FCW) using Dedicated Short-Range Wireless Communication (DSRC). In the cooperative collision warning technique, vehicles periodically broadcast the message in order to provide the driver with situational awareness and warning. In FCW, each vehicle is armed with localization and wireless communication device. Each vehicle periodically broadcasts a message about its current status to all neighboring vehicles in its transmission range. The neighboring vehicle receives the message and compares it with its own status for the computation of the likelihood of a collision. If the collision is imminent, the driver is advised to take an action. QualNet is the simulation tool used. Simulation results were battered toward characterizing the packet success probability trends with distance, uncovering a trade-off related to optimizing the application broadcast rate, and exploring performance trends with varying transmission ranges.

Misener et al. [12] proposed a cooperative collision warning between one vehicle to another vehicle using wireless communication. Three types of safety applications were demonstrated at Richmond field station. They are forward collision warning assistant, intersection assistant, and situational aware assistant.

Aldunate et al. [13] implemented an early vehicle crash detection application using smartphone technology. Three different smartphone vendors were considered. The values of normal driving and crash situations are obtained from sensors of smartphones and their gross values are calculated to detect early vehicle accidents and notify the accident information via SMS.

10.3.5 ITS IN REDUCING THE TRAVELING TIME

Prakash et al. [14] proposed adaptive QoS-based smart routing for VANETs by means of an ant colony optimization algorithm to find a smart route in Coimbatore city. The identified smart route can reduce traveling time. IEEE 802.11p is used for communication in VANET. RSR placement algorithm is proposed which indicates each RSR is simulated at a distance of 250m. A Road Side Receiver (RSR) holds information about moving vehicles. A client path is established from source to destination. The smart route is identified based on the reverse ANT colony optimization algorithm. AQRV and ACO are used to find the optimized path for detecting traffic congestion. Depending on the congestion level, an alternate path will be detected.

Tashiro et al. [15] implemented an optimal merging control of connected vehicles to reduce traveling time without a traffic signal due to planned paths, positions, vehicles, and other factors. This system predicts the conflicts between vehicles at an intersection in advance and decides the optimum merging control pattern to avoid the conflicts. Cellular automaton models such as NaSch model for traffic simulations, QS (Quick-Start) model for predicting forward movement and SIS (Slow-Start) model consider the inertia effect that temporarily stopped vehicles are delayed in acceleration. Four sets of local rules such as vehicle generation rules, speed rules, lane change rules, and link transfer rules were implemented. The merging control method is implemented by means of considering the following parameters such as lane change, acceleration, speed maintaining, deceleration, and stopping on its own.

Ko and Vaidya [16] proposed two Location Aided Routing (LAR) protocols which decrease the overhead in identifying routes. The LAR protocol uses information about the location from mobile hosts. Expected zone and request zone are the paradigms utilized at the time of route discovery which significantly reduces the routing overhead.

10.4   VANET PROTOCOLS AND APPLICATIONS

10.4.1   VANET PROTOCOLS

VANET applications can use any wireless technology as the basis for communication.

Some short-range radio technologies are ZigBee, Wi-Fi, and Bluetooth which are used for short-distance communication. Long-distance communication can use WiMax. In addition to that, cellular technologies such as LTE can also be used for communication.

Problems caused by the 802.11 DCF protocol were addressed and Propositional Fairness (PF)–based resource allocation on V2I architecture to overcome the problem was implemented by Harigovindan et al. [17]. PF resource allocation improves aggregate data transfer on both the single-lane and multi-lane on the V2I network.

Rahman and Tepe [18] proposed a cross-layer algorithm for V2I and V2V communication. Cross-Layer Extended Sliding Frame Reservation Aloha (CESFRA) is the mechanism that helps to disseminate data to up to three hops i.e., 900m. CESFRA was implemented based on R-Aloha.

Jiang and Delgrossi [19] provided a detailed overview of IEEE 802.11p, WAVE standard, and DSRC spectrum allocation. These standards can be used in vehicular environments.

Eichler et al. [20] proposed an efficient framework for data dissemination on heavy loaded scenarios. In this approach, vehicles independently evaluate the benefit of the packet. Local order between messages and the appropriate medium access is considered before transmitting it. Through this, the Quality of Service (QoS) standard is improved.

Cunha et al. [21] provided a summary of protocols available in VANET. They discussed the protocol stack for VANETs in terms of:

- Physical layer
- MAC layer
- Network layer

In the physical layer, the information about DSRC was discussed. The pictorial representation of IEEE 1609 (WAVE) architecture was described. It shows the relationship between IEEE 802.11p MAC and physical layers.

A comparison of MAC protocols along with its feature, medium of access, advantages, and drawbacks was represented under the MAC layer category. IEEE 802.11p can be used with Carrier-Sense Multiple Access with Collision Avoidance (CSMA/CA) access. DMAC can be used with Carrier-Sense Multiple Access (CSMA) access. ADHOC MAC and VC-MAC are based on ALOHA.

In the network layer, different communication paradigms such as unicast, multicast/geocast and broadcast communications were discussed and the comparison of routing protocols in VANET was given. Greedy Perimeter Stateless Routing (GPSR), GPCR, GPRSJ+, VADD, A-STAR, CAR, GyTAR, and MURU protocols can be used for unicast communication in V2V architectures. The PROMPT protocol is used for unicast communication in V2I architecture. IVG, direct message, and caching Geocast are the protocols used for Geocast communication with V2V architecture. BROAD COMM, UMB, V-TRADE/HV-TRADE, HyDi, and VoV are the protocols used for broadcast communication in V2V architectures. VANET applications are categorized into safety, efficiency, comfort, interactive environment, and urban sensing application. Challenges and future perceptions of VANET were also conferred.

Song et al. [22] presented a novel idea for a cluster-based routing protocol meant for deserts. They proposed an algorithm for establishing a cluster structure and cluster head election. Vehicle equipment priority, vehicle velocity priority, and overall election priority procedures were defined. Finally, the cluster-based routing protocol was checked against packet delivery ratio, routing cost, and end to end delay performance metrics.

Eiza et al. [23] presented a survey on QoS-aware routing protocols in an Ad hoc network along with its advantage and disadvantage. The protocols such as AODV-R, SAMQ, PBR, MABC, CAR, AQRV, GVGRID, and ARP-QD were discussed.

Eiza and Ni [24] implemented a graph theory–based VANET model. They proposed the VoEG model (VANET-Oriented Evolving Graph). It is based on graph theory. EG-Dijkstra algorithm was established to identify the MRJ (Most Reliable Journey) in the VoEG. They designed EG-RAODV routing protocol (Evolving Graph-Reliable Ad hoc On-Demand Distance Vector) for the routing scheme. EG-RAODV has been compared with reactive, proactive, and PBR protocols.

10.4.2 DENSITY ESTIMATION USING VANET

Panse et al. [25] introduced the concept for the detection of congestion on road and communicates the congestion information to other vehicles by Weighted Cluster Algorithm (WCA) which improves the performance of the network in complicated scenarios. Road Side Units (RSUs) are linked by high-speed buses through underground cables. RSUs collect information on moving vehicles and communicate the same with the server. The decisional server gets the information from the RSUs and communicates the traffic and accident information over the highway among vehicles. The provided information is shared among Vehicles to Vehicles (V2V) by the WCA algorithm.

Akhtar et al. [26] implemented three density estimation distributed algorithms, i.e., sample and collide, hop sampling, and gossip-based aggregation. These algorithms are implemented at different vehicle traffic densities in both highways and urban areas. Among the three algorithms, hop sampling delivers the highest accuracy in the least convergence time and introduces the least overhead on the network. The sample and collide algorithm is based on an equal sampling of the nodes from the population and then determining the density, based on the number of samples of nodes that are collected, before an already sampled node is re-selected. Hop sampling is implemented based on the principle of probabilistic polling. The initiator sends a message to all the nodes by gossiping. The nodes respond back to the initiator probabilistically based on their distance from it. Depending on the replies obtained from other nodes to the initiator node, the size of the network is estimated. In gossip-based aggregation, each peer periodically exchanges information with one of its neighbors chosen at random to find the size of the network.

10.4.3   VANET APPLICATIONS

Kumar et al. [27] categorized various possible applications in VANET. Four types of applications are described:

- Safety-oriented application
- Commercial-oriented application
- Convenience-oriented application
- Productive application

Safety applications include real-time traffic, cooperative message transfer, post-crash notification, road hazard control notification, Cooperative Collision Warning, and Traffic Vigilance.

Commercial applications are remote vehicle personalization/diagnostics, value-added advertisement, Internet access, digital map downloading, and real-time video relay.

Route diversions, electronic toll collection, parking availability, and active prediction are convenience applications.

Environmental benefits, time utilization, and fuel-saving related applications fall under the productive application category.

Syfullah and Lim [28] implemented a system by combining the traditional 802.11p standard VANET networks with LTE networks to form a hybrid cloud-VANET. The cloud-VANET architecture is depicted by four levels namely Vehicular Cloud Network (VCN), Digital Content Network (DCN), Infrastructure Cloud Network (ICN) and Server-to-Cloud Network (S2CN). Cloud leader starts the cloud-VANET. Three algorithms such as IEEE 802.11p-LTE cloud member algorithm, cloud leader algorithm, and eNodeB algorithm are proposed. These algorithms reduce the broadcast storm issues by decreasing repeated data broadcasting and keeping overheads low. This approach facilitates the VANET to utilize high bandwidth along with uninterruptable data connectivity. SUMO and OMNET++ are used as simulation tools and simulated for highway scenarios. Data packet loss and SNIR loss were reduced by about 30% when compared with IEEE802.11p and the reference scheme.

Hussain and Oh [29] provided a co-operation aware, VANET-based cloud computing. Co-operation as a Service (CaaS) defines the co-operation of moving cars on the road which is further divided into Traffic Information as a Service (TIaaS) and Warning as a Service (WaaS). TIaaS provides the traffic flow information to the subscribers by cloud decision module. WaaS provides subscribers with warning information. Gateway terminals are used for communication between the cloud and vehicles and vice versa.

Jose et al. [30] proposed intelligent vehicle monitoring using GPS and cloud computing. Each vehicle is registered by the vehicle owner in a cloud server using a web portal which generates a vehicle ID. GPS is used to collect vehicle information such as current location, distance covered by vehicle, and prediction of arrival time and the data are transmitted to the cloud server via satellite communication. The sensors such as fuel-level sensors and breath alcohol sensors are used to find fuel-related information in vehicles and the influence of alcohol on drivers. The sensed data are transferred to the server through GSM with the nearest access point using ZigBee IEEE 802.15.4. The best-fit algorithm helps to store the data in the server. From the cloud server, the vehicle owner can monitor the vehicle.

Garai et al. [31] presented a survey on vehicular cloud challenges and their opportunities. Fundamentals of vehicular cloud networking along with advantages, possible applications, and challenges were discussed.

Yu et al. [32] integrated VANET with cloud computing. Resources such as computation resources, bandwidth, and storage resources are shared by vehicles. Optimal cloud resource allocation is done by the game theoretical approach. The cloud types include the vehicular, roadside, and cellular cloud. Resource reservation scheme is used to solve the problem of virtual machine migration.

10.4.4   VANET Simulators

Mobility simulators, network simulators, and vehicular network simulators are the types of simulators used in the implementation of VANET applications.

10.4.4.1   Mobility Simulators

Mobility simulators are used to generate the map along with vehicles. "SUMO" is an open-source command line simulation. It is used to simulate road traffic. SUMO was used in [26, 28]. "VantMobiSim" is an agent-based microscopic and macroscopic mobility simulation software used in [17]. "CARISMA" is a traffic simulator developed by BMW and it was used in [20]. FreeSim, NetSim, VISSIM, and PARAMICS are the other type of mobility simulator which are available.

10.4.4.2   Network Simulators

Network simulators allow us to specify the communication type and its protocols for implementation. Some of the network simulators are described below.

"OMNET++" is a component-based C++ simulation library to build network-ing scenarios. [23, 24, 28] used OMNET++. "Nam" is an animation tool to view the network simulation and it is used in [25]. "NCTUns" is a simulator as well as an emulator used in [1]. "NS2" is an open-source simulator written in C++ and Python used in [14, 17]. "QualNet" is a planning, testing and training tool for network communication used in [11]. "SWANS" is a tool which is built on top of Java in simulation time and it was used in [5]. "MaRs" is a Maryland Routing Simulator used in [16]. Some other network simulators are J-SIM, SNS and JiSt/SWA. STRAW and JiSt/SWA can be directly connected without any vehicular network simulator.

10.4.4.3   Vehicular Network Simulators

VANET simulators act as an intermediate between mobility and network simulator. Some VANET simulators are discussed below.

"GrooveNet" can integrate a mobility simulator and a network simulator. The study [10] was implemented using a Groovenet simulator. "Veins" can integrate SUMO and OMNET++. "MOVE" integrates SUMO and NS2 as well as SUMO and QualNet. "TraNs" integrates SUMO and NS2. "MobiReal" can connect "GTNetS" (network simulator) and NetStream (mobility simulator).

By using these simulators, any type of VANET application can be implemented for a specific need.

10.5   COMPARISON OF "VANET" APPLICATIONS AND PROTOCOLS THAT AID "ITS"

Table 10.1 illustrates "VANET architecture its applications, implementation tech-niques, and performance metrics".

10.6   CONCLUSION

This chapter provides an overview of the literature which includes various VANET applications such as traffic management in ITS, accident avoidance, and ITS in reducing traveling time. Communication and routing protocols of VANET were dis-cussed. The VANET simulators used for the implementation were given. The com-parison of already existing systems was summarized in Table 10.1. ITS and VANET combines the accuracy and efficiency of road transportation and so many human lives can be saved.

TABLE 10.1
Comparision of "VANET" Applications and Protocols That Aid "ITS"

Literature	VANET Architecture Used	Implemented Application	Proposed Methodology	Communication and Routing Protocols	Performance Metrics
Maslekar et al.[1]	V2V V2I	VANET-based adaptive traffic signal control	Density-based cluster using VANET (DBCV) algorithm is used for density of vehicle estimation. Webster's formula is modified and used for calculating cycle time and optimum length	Inter-cluster communication	Number of cars approaching the traffic signal intersection is compared with the average waiting time and concluded that VANET-based traffic signal control reduces the number of vehicles stopping at intersection is reduced by 20 percent
Watanabe et al. [3]	V2V	Dynamic Map 2.0 use case considered to test the dyanamic map is cooperative merging assistance	Dynamic map 2.0 prototype is developed using SQL query-based approach with the concept of sliding window	GPRS is used for cellular communication	In-vehicle information, Local area information, City-scale information are considered for the scope of data management
Sam et al.[10]	I2V	Pedestrian safety	GPS is used to find the location of pedestrians and send them to a nearby RSU. RSU is used to alert the motorist about the pedestrian location	Wi-Fi (IEEE 802.11)	Pedestrian speed and Car speed are considered to analyze the accident possibility
ElBatt et al. [11]	V2I	Cooperative collision warning (CCW)	Forward collision technique using dedicated short-range communication is implemented	DSRC (IEEE 802.11p)	Packet inter-reception time, cumulative number of packet reception, packet success probability, per-packet latency are considered to quantify the communication performance of CCW applications

(Continued)

TABLE 10.1 (CONTINUED)
Comparision of "VANET" Applications and Protocols That Aid "ITS"

Literature	VANET Architecture Used	Implemented Application	Proposed Methodology	Communication and Routing Protocols	Performance Metrics
Prakash et al.[14]	V2V V2I	Smart route for vehicles	Road Side Receiver (RSR) placement algorithm to decide the number of roadside units required. Reverse ANT colony optimization algorithm to find the best route	DSRC (IEEE 802.11p)	Connectivity, packet success probability, packet delivery ratio, and delay are considered to show the improved probability of their proposed system
Harigovindan et al.[17]	V2I	Proportional fair (PF) resource allocation	Data transfer rates on single-lane and multi-lane are computed	IEEE 802.11b (EDCA)	Data transfer rates across different zones are considered to overcome the unfair resource allocation
Panse et al. [25]	V2V V2I Adhoc Network	Load detection on highways	Weighted cluster algorithm to improve the performance of the network and used for communication between vehicle-to-vehicle (V2V)	Wi-Fi AODV	Connectivity, mobility, throughput, packet delivery ratio, routing overhead, end to end delay are considered to quantify the proposed system
Akhtar et al. [26]	V2V	Density estimation of vehicles in a distributed environment	Sample and collide algorithm, hop sampling algorithm, and gossip-based aggregation are implemented	IEEE 802.11p	Density estimation, convergence time, overhead, error ratio, load on initiator are the metrics considered to find the best among the three algorithms
Syfullah and Lim [28]	V2V V2I I2Cloud	VANET – cloud integration	Cloud leader algorithm, cloud member algorithm, and eNodeB algorithm are proposed	LTE IEEE 802.11p	Packet loss and SNIR loss are considered to reduce the number of broadcasts among nodes

REFERENCES

1. Maslekar, N., Boussedjra, M., Mouzna, J., & Labiod, H. (2011, May). VANET based adaptive traffic signal control. In *2011 IEEE 73rd Vehicular Technology Conference (VTC Spring)* (pp. 1–5). IEEE.
2. Lykov, S., & Asakura, Y. (2018). Anomalous traffic pattern detection in large urban areas: Tensor-based approach with continuum modeling of traffic flow. *International Journal of Intelligent Transportation Systems Research*,18(1), 1–9.
3. Watanabe, Y., Sato, K., & Takada, H. (2018). DynamicMap 2.0: A traffic data management platform leveraging clouds, edges and embedded systems. *International Journal of Intelligent Transportation Systems Research*, 18(12), 1–13.
4. Mizuno, K., Fukui, Y., & Nishihara, S. (2008, January). Urban traffic signal control based on distributed constraint satisfaction. In *Proceedings of the 41st Annual Hawaii International Conference on System Sciences (HICSS 2008)* (pp. 65–65). IEEE.
5. Choffnes, D. R., & Bustamante, F. E. (2005). *Straw-an Integrated Mobility and Traffic Model for Vanets*. Northwestern Univ Evanston IL DEPT of Computer Science.
6. Osafune, T., Takahashi, T., Kiyama, N., Sobue, T., Yamaguchi, H., & Higashino, T. (2017). Analysis of accident risks from driving behaviors. *International Journal of Intelligent Transportation Systems Research*, 15(3), 192–202.
7. Devi, B., Bavatharini, S. S., Samyuktha, G., Shobica, S., & Sonia, E. (2018, March). Voice alert for accident avoidance on merging lanes, blind curves and T junctions. In *2018 Second International Conference on Electronics, Communication and Aerospace Technology (ICECA)* (pp. 418–422). IEEE.
8. Enriquez, D., Jenson, S., Bautista, A., Hawn, P., Kim, S. I., Ali, M., & Miller, J. (2017). On software-based remote vehicle monitoring for detection and mapping of slippery road sections. *International Journal of Intelligent Transportation Systems Research*, 15(3), 141–154.
9. Bhumkar, S. P., Deotare, V. V., & Babar, R. V. (2012). Accident avoidance and detection on highways. *International Journal of Engineering Trends and Technology*, 3(2), 247–252.
10. Sam, D., Evangelin, E., & Raj, V. C. (2015, February). A novel idea to improve pedestrian safety in Black Spots using a hybrid VANET of vehicular and body sensors. In *International Conference on Innovation Information in Computing Technologies* (pp. 1–6). IEEE.
11. ElBatt, T., Goel, S. K., Holland, G., Krishnan, H., & Parikh, J. (2006, September). Cooperative collision warning using dedicated short range wireless communications. In *Proceedings of the 3rd International Workshop on Vehicular Ad Hoc Networks* (pp. 1–9). ACM.
12. Misener, J. A., Sengupta, R., & Krishnan, H. (2005, November). Cooperative collision warning: Enabling crash avoidance with wireless technology. In *12th World Congress on ITS* (pp 1–11, Vol. 3).
13. Aldunate, R. G., Herrera, O. A., & Cordero, J. P. (2013). Early vehicle accident detection and notification based on smartphone technology. In *Ubiquitous Computing and Ambient Intelligence: Context-Awareness and Context-Driven Interaction: 7th International Conference, UCAmI 2013, Carrillo, Costa Rica, December 2-6, 2013, Proceedings* (pp. 358–365). Springer International Publishing.
14. Prakash, J., Sengottaiyan, N., & Nandhini, S. H. (2018). Smart routing for vehicle at optimal position with ant colony optimization and AQRV in VANET. *EAI Endorsed Transactions on Energy Web*, 5(20), e10.
15. Tashiro, M., Motoyama, H., Ichioka, Y., Miwa, T. and Morikawa, T., (2020). Simulation analysis on optimal merging control of connected vehicles for minimizing travel time. *International Journal of Intelligent Transportation Systems Research, 18,*

pp.65-76.Tashiro, M., Motoyama, H., Ichioka, Y., Miwa, T., & Morikawa, T. (2018). Simulation analysis on optimal merging control of connected vehicles for minimizing travel time. *International Journal of Intelligent Transportation Systems Research*, 18, 65–76.

16. Ko, Y. B., & Vaidya, N. H. (2000). Location-Aided Routing (LAR) in mobile ad hoc networks. *Wireless Networks*, 6(4), 307–321.
17. Harigovindan, V. P., Babu, A. V., & Jacob, L. (2014). Proportional fair resource allocation in vehicle-to-infrastructure networks for drive-thru internet applications. *Computer Communications*, 40, 33–50.
18. Rahman, K. A., & Tepe, K. E. (2014, June). Towards a cross-layer based MAC for smooth V2V and V2I communications for safety applications in DSRC/WAVE based systems. In *2014 IEEE Intelligent Vehicles Symposium, Proceedings* (pp. 969–973). IEEE.
19. Jiang, D., & Delgrossi, L. (2008, May). IEEE 802.11 p: Towards an international standard for wireless access in vehicular environments. In *VTC Spring 2008-IEEE Vehicular Technology Conference* (pp. 2036–2040). IEEE.
20. Eichler, S., Schroth, C., Kosch, T., & Strassberger, M. (2006, July). Strategies for context-adaptive message dissemination in vehicular ad hoc networks. In *2006 Third Annual International Conference on Mobile and Ubiquitous Systems: Networking & Services* (pp. 1–9). IEEE.
21. Cunha, F., Villas, L., Boukerche, A., Maia, G., Viana, A., Mini, R. A., & Loureiro, A. A. (2016). Data communication in VANETs: Protocols, applications and challenges. *Ad Hoc Networks*, 44, 90–103.
22. Song, T., Xia, W., Song, T., & Shen, L. (2010, November). A cluster-based directional routing protocol in VANET. In *2010 IEEE 12th International Conference on Communication Technology* (pp. 1172–1175). IEEE.
23. Eiza, M. H., Owens, T., Ni, Q., & Shi, Q. (2015). Situation-aware QoS routing algorithm for vehicular ad hoc networks. *IEEE Transactions on Vehicular Technology*, 64(12), 5520–5535.
24. Eiza, M. H., & Ni, Q. (2013). An evolving graph-based reliable routing scheme for VANETs. *IEEE Transactions on Vehicular Technology*, 62(4), 1493–1504.
25. Panse, P., Shrimali, T., & Dave, M. (2016). An approach for preventing accidents and traffic load detection on highways using V2V communication in VANET. *International Journal of Information, Communication and Computing Technology (IJICCT)*, 4(1), 181–186.
26. Akhtar, N., Ergen, S. C., & Ozkasap, O. (2012, November). Analysis of distributed algorithms for density estimation in vanets (poster). In *2012 IEEE Vehicular Networking Conference (VNC)* (pp. 157–164). IEEE.
27. Kumar, V., Mishra, S., & Chand, N. (2013). Applications of VANETs: Present & future. *Communications and Network*, 5(1), 12.
28. Syfullah, M., & Lim, J. M. Y. (2017, February). Data broadcasting on cloud-VANET for IEEE 802.11 p and LTE hybrid VANET architectures. In *2017 3rd International Conference on Computational Intelligence & Communication Technology (CICT)* (pp. 1–6). IEEE.
29. Hussain, R., & Oh, H. (2014). Cooperation-aware VANET clouds: Providing secure cloud services to vehicular ad hoc networks. *JIPS*, 10(1), 103–118.
30. Jose, D., Prasad, S., & Sridhar, V. G. (2015). Intelligent vehicle monitoring using global positioning system and cloud computing. *Procedia Computer Science*, 50, 440–446.
31. Garai, M., Rekhis, S., & Boudriga, N. (2015, July). Communication as a service for cloud VANETs. In *2015 IEEE Symposium on Computers and Communication (ISCC)* (pp. 371–377). IEEE.
32. Yu, R., Zhang, Y., Gjessing, S., Xia, W., & Yang, K. (2013). Toward cloud-based vehicular networks with efficient resource management. *IEEE Network*, 27(5), 48–55. arXiv Preprint ArXiv:1308.6208

11 Observer-Based Control for Rotary Servo System

Arun Kumar Pinagapani,
Praveen Kumar Mageswaran,
Tamilselvan Arunachalam, S. P. Naveen, and
V. Gokul Nitheesh

CONTENTS

11.1 INTRODUCTION

A normal pendulum is stable when hanging downwards, an inverted pendulum is an underactuated mechanical system which is inherently unstable. A rotary inverted pendulum (RIP) system is set up by using a rotary encoder, servo motor, and some mechanical structures. The rotary inverted pendulum is a benchmark system stabilized using numerous different control algorithms and is developed from a linear inverted pendulum. The equations of motion of a RIP are non-linear. In the area of control issues related to an inverted pendulum system, many studies are found. Ding, Z. and Li, Z. A [1] proposed a cascade fuzzy control system for inverted pendulum

DOI: 10.1201/9781003307716-11

based on the Mamdani–Sugeno type. Ogata, K [2] discussed a comprehensive study of controllers. Mehmet Öksüz et al. [3] presented an alternative controller design for rotary inverted pendulum (RIP). Gou Yun Tao et al. [4] focused on the designing and realization of a RIP based on STM32. Velchuri Srisha et al. [5] have described a comparative study of controllers for stabilizing a rotary inverted pendulum. Wen-Hua Chen et al. [6] presented an overview of a disturbance observer-based (DOB) control and related methods. Abhishek Kathpal et al. [7] have dealt with simulation and real-time control of RIP and SimMechanics™ based modeling. Akhtaruzzaman Md et al. [8] focused on the control of a RIP using various methods, comparative assessment, and result analysis. Francesco Ferrante et al. [9] have proposed a solution of observer-based control for linear systems with quantized output. George Ellis has presented a practical guide for the observer in control system. Vonkomer J. et al. [10] developed a disturbance observer control for AC speed servo with improved noise attenuation. Chen Xi-Song et al. [11] presented a disturbance observer enhanced PID decoupling control for multi-variable processes. Quyen et al. [12] focused on RIP and control of rotary inverted pendulum by Artificial Neural Network (ANN). This chapter presents a non-linear and linearized model of the RIP, four full-state controllers are designed by using MATLAB for the system stabilizing and to obtain a linear dynamic model. The results of the simulation of a controller with different damping ratios are compared and hence the best result is taken into the observer design to a RIP system.

11.2   SYSTEM MODELING

The mathematical model is often used to analyze the system and provides an easy way to simulate the model in software. The system modeling gives the derivation of the mathematical model of the rotary inverted pendulum.

11.2.1   MATHEMATICAL MODEL OF ROTARY INVERTED PENDULUM

Two generalized coordinates for this problem are α and θ which are independent. T is the sum of the rotary arm's kinetic energy T_{arm} and the pendulum's kinetic energy T_{pend} are given in Equation (11.1). V is the sum of the rotary inverted pendulum potential energy, which includes only the pendulum's potential energy, and it is given in Equation (11.2).

The rotary inverted pendulum in [3] is used in this work.

$$T = T_{pend} + T_{arm} = \frac{1}{2}m_p\dot{P}^2 + \frac{1}{2}J_p\dot{\alpha}^2 + \frac{1}{2}J_r\dot{\theta}^2 \tag{11.1}$$

$$V = \frac{1}{2}m_PgL_p\cos\alpha \tag{11.2}$$

To attain the velocity of the center of the pendulum, the coordinates of the center of mass of the pendulum are determined in Equations (11.3) to (11.5).

$$P_x = L_r \cos\theta + \frac{1}{2} L_p \sin\alpha \sin\theta \tag{11.3}$$

$$P_y = L_r \sin\theta - \frac{1}{2} L_p \sin\alpha \cos\theta \tag{11.4}$$

$$P_z = \frac{1}{2} L_p \cos\alpha \tag{11.5}$$

By taking derivatives of the positions, the velocity of point P is found. The Lagrange equation becomes Equation (11.6).

$$L = \begin{cases} \dfrac{1}{2}\left(J_r + m_p L_r{}^2 + \dfrac{1}{4} m_p L_p{}^2 \sin^2\alpha\right)\dot{\theta}^2 \\[2mm] \dot{\theta}^2 + \dfrac{1}{2}\left(J_p + \dfrac{1}{4} m_p L_p{}^2\right)\dot{\alpha}^2 - \\[2mm] \dfrac{1}{2} m_p L_r L_p \cos\alpha\,\dot{\theta}\dot{\alpha} - \\[2mm] \dfrac{1}{2} m_p g L_p \cos\alpha \end{cases} \tag{11.6}$$

The system dynamic model results in Equations (11.7) and (11.8). Here T is the applied torque, B_r is the viscous friction coefficient of the torque and B_p is the viscous damping coefficient of the pendulum.

Torque based on the rotary arm is generated by a servo motor. It is defined in Equation (11.9) where η_g, η_m, K_g, R_m, k_t, and V_m are the efficiency of the gear, the efficiency of the motor, the gear ratio, the motor armature resistance, the motor current, the torque constant, and the input motor voltage.

$$\left. \begin{aligned} \ddot{\theta}\left(J_r + m_p L_r{}^2 + \frac{1}{4} m_p L_p{}^2 \sin^2\alpha\right) + \\[1mm] \ddot{\alpha}\left(-\frac{1}{2} m_p L_r L_p \cos\alpha\right) + \\[1mm] \dot{\theta}\dot{\alpha}\left(\frac{1}{2} m_p L_p{}^2 \sin\alpha\cos\alpha\right) + \dot{\alpha}^2\left(\frac{1}{2} m_p L_r L_p \sin\alpha\right) \end{aligned} \right\} = \tau - B_r\dot{\theta} \tag{11.7}$$

$$\left. \begin{aligned} \left(-\frac{1}{2} m_p L_r L_p \cos\alpha\right) + \ddot{\alpha}\left(J_p + \frac{1}{4} m_p L_p{}^2\right) + \\[1mm] \dot{\theta}^2\left(-\frac{1}{4} m_p L_p{}^2 \sin\alpha\cos\alpha\right) - \\[1mm] \frac{1}{2} m_p g L_p \sin\alpha \end{aligned} \right\} = -B_p\dot{\alpha} \tag{11.8}$$

$$\tau = \frac{\eta_g \eta_m K_g K_t \left(V_m - K_g K_m \dot{\theta} \right)}{R_m} \tag{11.9}$$

For a generalized coordinate, vector q, can be generalized into the matrix form as Equation (11.10) in which D is the inertial matrix, C is the damping matrix and $g(q)$ is the gravitational vector.

$$\ddot{q} D(q) + \dot{q} C(q, \dot{q}) + g(q) = \tau \tag{11.10}$$

Equations (11.7) and (11.8) are non-linear equations. They are to be linearized at the zero initial conditions as $\left[\theta \; \alpha \; \dot{\theta} \; \dot{\alpha} \; \ddot{\theta} \; \ddot{\alpha} \right] = \left[000000 \right]$. The resultant mathematical model of the inverted pendulum is defined as matrix form in Equation (11.11).

$$\begin{bmatrix} \tau \\ 0 \end{bmatrix} = \begin{bmatrix} J_r + m_p L_r^2 & -\dfrac{1}{2} m_p L_r L_p \\ -\dfrac{1}{2} m_p L_r L_p & J_p + \dfrac{1}{4} m_p L_p^2 \end{bmatrix} \begin{bmatrix} \ddot{\theta} \\ \ddot{\alpha} \end{bmatrix} + \begin{bmatrix} B_r & 0 \\ 0 & B_p \end{bmatrix} \begin{bmatrix} \dot{\theta} \\ \dot{\alpha} \end{bmatrix} + \begin{bmatrix} 0 & 0 \\ 0 & -\dfrac{1}{2} m_p g L_p \end{bmatrix} \begin{bmatrix} \theta \\ \alpha \end{bmatrix}$$

$$\tag{11.11}$$

The equation for the angular acceleration of the arm θ, and the angular acceleration of the pendulum α should be determined by using Equation (11.11). The equations can be simplified by defining a constant gain G in Equation (11.12).

$$G = \frac{1}{\left(J_r + m_p L_r^2 \right) \left(J_p + \dfrac{1}{4} m_p L_p^2 \right) - \dfrac{1}{4} m_p^2 L_r^2 L_p^2} \tag{11.12}$$

The rotary inverted pendulum has two coordinates for description. Two second-order equations; thus it has four state variables where: $X_1 = \theta; X_2 = \alpha; X_3 = \dot{\theta}$ and $X_4 = \dot{\alpha}$. The general state space equations are given in Equation (11.13).

$$\dot{x} = Ax + Bu \text{ and } y = Cx + D \tag{11.13}$$

11.2.2 System Parameters

The mathematical model for the system firmly depends on the physical parameters of a rotary inverted pendulum such as pendulum mass (m_p), arm full length (L_r),

TABLE 11.1

Technical Specifications of RIP

Parameters	Symbols	Values	Units
Gearbox efficiency	η_g	0.9	–
Motor efficiency	η_m	0.69	–
Motor torque constant	k_t	0.007683	$N.m / A$
Total gear ratio	K_g	70	–
Motor back-EMF constant	k_m	0.0076777	$V.s / rad$
Motor armature resistance	R_m	2.6	$\&$
Pendulum mass	m_p	0.127	kg
Arm full length	L_r	0.2159	m
Pendulum full length	L_p	0.33655	m
Arm moment of inertia	J_r	0.0009983	$kg.m^2$
Pendulum moment of inertia	J_p	0.0012	$kg.m^2$
Equivalent arm viscous damping coefficient	B_r	0.0024	$N.m.s / rad$
Equivalent Pendulum viscous damping coefficient	B_p	0.0024	$N.m.s. / rad$
Gravity	g	9.81	$kg.m$

pendulum full length (L_p), arm moment of inertia (J_p) and its appropriate parameters that are shown in Table 11.1, whereby state space model is obtained. Furthermore, the state space model is used while performing simulations with respective parameters [2].

Matrices A, B, C, and D are formed using parameter values. A is the state matrix that gives information about the characteristic of the system which are shown in Equation (11.14); B is the input-to-state matrix which is given in Equation (11.16); C is the state-to-output matrix which is given in Equation (11.18); and D is the feed through the matrix which is given in Equation (11.19).

$$A = \begin{bmatrix} 0 & 0 & 1 & 0 \\ 0 & 0 & 0 & 1 \\ 0 & Y_1 & Y_2 & Y_3 \\ 0 & Y_4 & Y_5 & Y_6 \end{bmatrix} \tag{11.14}$$

$$Y_1 = \frac{1}{4} Gm^2 gL_r L_p^2 \ , Y_2 = -GB_r \left(J_p + \frac{1}{4} mL_p^2 \right)$$

$$Y_3 = -\frac{1}{2}GmL_pL_rB_r \ ,Y_4 = \frac{1}{2}GmgL_p\left(J_r + mL_p^2\right)$$

$$Y_5 = -\frac{1}{2}GmL_pL_rB_r \ ,Y_6 = -GB_p\left(J_r + \frac{1}{4}mL_r^2\right)$$

$$A = \begin{bmatrix} 0 & 0 & 1 & 0 \\ 0 & 0 & 0 & 1 \\ 0 & 81.34 & -28.81 & -0.93 \\ 0 & 121.96 & -28.18 & -1.40 \end{bmatrix} \qquad (11.15)$$

$$B = \begin{bmatrix} 0 \\ 0 \\ G\left(J_p + \frac{1}{4}mL_p^2\right) \\ \frac{1}{2}GmL_pL_r \end{bmatrix} \qquad (11.16)$$

$$B = \begin{bmatrix} 0 \\ 0 \\ 51.81 \\ 49.84 \end{bmatrix} \qquad (11.17)$$

$$C = \begin{bmatrix} 1 & 0 & 0 & 0 \\ 0 & 1 & 0 & 0 \end{bmatrix} \qquad (11.18)$$

$$D = \begin{bmatrix} 0 \\ 0 \end{bmatrix} \qquad (11.19)$$

11.3   IMPLEMENTATION

The inverted pendulum is an unstable state because its center of gravity is over its axis of rotation. It is stable when it directs downwards. The objective is to hold the pendulum at its upright position by using the controller, to enhance its performance by adding the observer to the controller.

11.3.1   DESIGN OF THE CONTROLLER BASED ON POLE PLACEMENT

A linear dynamic system in the state space form is given in Equation (11.20) where $D = 0$. To stabilize the system to improve its response, full-state feedback is $u = -Kx$.
$\dot{x} = Ax + Bu$ and

$$y = Cx + Du \qquad (11.20)$$

The closed-loop system is given in Equation (11.21). Stabilization of the system is obtained using state feedback control. Thus, all closed-loop poles should be located on the left-hand of the complex plane.

$\dot{x} = (A - BK)x$ and

$$y = Cx \qquad (11.21)$$

11.3.2 System Poles

The system poles give information on the system's characteristic. To define the poles of the system

$det(SI - A) = 0$ is estimated. The poles of the system are found as: $p_1 = 0$, $p_2 = -32.39$, $p_3 = 7.32$ and $p_4 = -5.15$. When the inverted pendulum is kept in an upright position, a small perturbation will remove the inverted pendulum from the stable position is tend to make the system unstable. The inverted pendulum will not return back to the upright position and it is not stable since one of the poles is in the right-hand plane. The pole, or poles, are transferred in the left-hand plane from the right-hand plane to provide the stability of the system [1].

11.3.3 Desired Poles

There are four eigenvalues of the system. Two of them are specified as follows: $p_1 = -\delta W_n + jW_d$ and $p_2 = -\delta W_n - jW_d$. They are the complex conjugate dominant poles [3], where δ is the damping ratio, W_n is the natural frequency, and the other poles p_3 and p_4 are defined at -10 and -20.

The natural frequency is taken $W_n = 4$ rad/s [2], and four values of damping ratios are used. The desired poles are found according to the damping ratios introduced. The required systems are given with their poles in Table.11.2.

11.3.4 Full-State Controller Coefficient – Pole Placement

The controllability of the system should be checked to decide whether the system can be controlled by a full-state controller or not. If A and B matrices of the state-space

TABLE 11.2
Desired Poles for Different Damping Ratios

No.	Damping ratio	p_1	p_2	p_3	p_4
1.	0.7	−2.8+2.86j	−2.8−2.86j	−10	−20
2.	0.75	−3+2.64j	−3−2.64j	−10	−20
3.	0.8	−3.2+2.4j	−3.2−2.4j	−10	−20
4.	0.85	−3.4+2.11j	−3.4−2.11j	−10	−20

model are controllable, the system can be controlled by the pole placement method with full-state control gains [2]. The controllability matrix of the system is determined by Equation (11.22).

$$M = \begin{bmatrix} B\, AB\, A^2 BA^3 B...A^n B \end{bmatrix}$$ (11.22)

The rank of M is four. The system is controllable with four states. The general equation form of the open loop characteristic equation is given in Equation (11.24) and companion matrices of A and B should be estimated in the z-plane. They can be denoted by A_z and B_z as given in Equation (11.25) and the controllability matrix of the companion M_z could be accounted for in Equation (11.23).

$$M_z = \begin{bmatrix} B_z A_z B_z A_z^2 B_z...A_z^n B_z \end{bmatrix}$$ (11.23)

$$s^4 + a_3 s^3 + a_2 s^2 + a_1 s + a_0$$ (11.24)

$$A_z = \begin{bmatrix} 0 & 1 & 0 & 0 \\ 0 & 0 & 1 & 0 \\ 0 & 0 & 0 & 1 \\ -a_0 & -a_1 & -a_2 & -a_3 \end{bmatrix}, B_Z = \begin{bmatrix} 0 \\ 0 \\ 0 \\ 1 \end{bmatrix}$$ (11.25)

Control gain in the z-plane, K_z, should be computed to assign the poles of $A_z -$ $B_z K_z$ to the required places. The closed-loop equation form for desired poles is given in Equation (11.26) and $K_z = \begin{bmatrix} K_{z1} K_{z2} K_{z3} K_{z4} \end{bmatrix}$. Calculations are given in Equation (11.27).

$$(s - p_1)(s - p_2)(s - p_3)(s - p_4) = s^4 + a_{d3} s^3 + a_{d2} s^2 + a_{d1} s + a_{d0}$$ (11.26)

$$\begin{bmatrix} 0 & 1 & 0 & 0 \\ 0 & 0 & 1 & 0 \\ 0 & 0 & 0 & 1 \\ -a_0 - K_{z1} & -a_1 - K_{z2} & -a_2 - K_{z3} & -a_3 - K_{z4} \end{bmatrix} = \begin{bmatrix} 0 & 1 & 0 & 0 \\ 0 & 0 & 1 & 0 \\ 0 & 0 & 0 & 1 \\ -a_{d0} & -a_{d1} & -a_{d2} & -a_{d3} \end{bmatrix}$$

$$K_Z^T = \begin{bmatrix} a_{d0} - a_0 \\ a_{d1} - a_1 \\ a_{d2} - a_2 \\ a_{d3} - a_3 \end{bmatrix}$$ (11.27)

TABLE 11.3

Controllers for Different Damping Ratios

Damping ratio	K_1	K_2	K_3	K_4
0.7	−1.4147	12.8302	−1.3991	1.5625
0.75	−1.4103	13.1079	−1.4366	1.6096
0.8	−1.4130	13.3959	−1.4755	1.6580
0.85	−1.4140	13.6816	−1.5140	1.7061

The transformation matrix, Z should be calculated by multiplication of controllability and inverse controllability of the companion, $Z = M.M_Z^{-1}$. Finally, the control gain from the z-plane should be transformed into actual plane control gain. The system poles are located in the desired poles by estimating the control gain, $K = K_z Z^{-1} = \begin{bmatrix} K_1 K_2 K_3 K_4 \end{bmatrix}$, where K is the control gain. The control gains for using different damping ratios are given in Table 11.3.

11.3.5 SIMULATION OF A CONTROLLER

In Figure 11.1, the simulation setup is given, where the output θ shows the angular position of the arm, α shows the angular position of the pendulum. A step input of magnitude 10 degrees is given as an input. All controller gains are shown in Table 11.3.

11.3.6 STATE OBSERVER

The estimation of the internal state of a given real-time system is provided by a state observer, from measurements of the input and output of the real system. The observer is deployed where no need to measure all the states of the system to estimate the

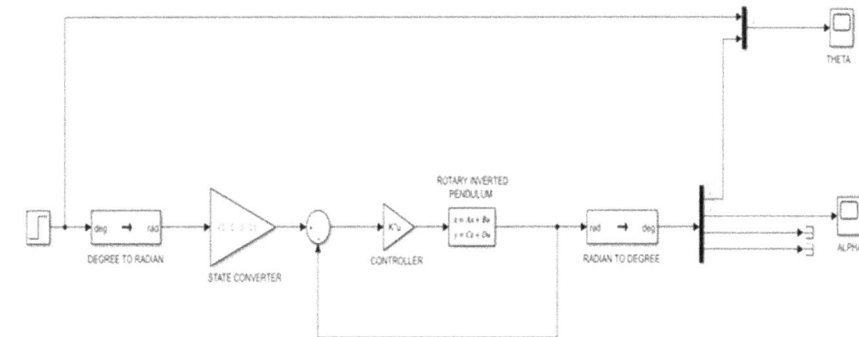

FIGURE 11.1 Simulation of the controller

unmeasured states from the outputs. It is typically computer-implemented, whereby the basis of the many practical systems is provided. The system state is important to solve many control theory problems; for instance, stabilizing a system using state feedback. Sometimes direct observation is not sufficient to get the physical state of the system in most practical cases. Instead, system outputs provide the indirect effects of the internal state. It is possible to fully reconstruct the system state from its output measurements using the state observer when the system can be observable.

11.3.7 CONTINUOUS-TIME OBSERVER MODEL

The observer gains are chosen to make the continuous-time error dynamics converge to zero asymptotically (i.e., when $A - LC$ is a Hurwitz matrix). The observer model of the physical system is then typically derived from Equation (11.20) Additional terms may be included in order to ensure that, the model's state converges to that of the plant receiving successive measured values of the plant's inputs and outputs. In particular, the output of the observer may be subtracted from the output of the plant and then multiplied by a matrix L [13]; then it is substituted in the equations for the state of the observer to generate a so-called Luenberger observer, defined by:

$$\dot{\hat{x}} = A\hat{x} + Bu + L\left(y - \hat{y}\right) \tag{11.28}$$

$$\hat{y} = C\hat{x} + Du \tag{11.29}$$

The observer error $e = x - \hat{x}$, satisfies the equation:

$$\dot{e} = \left(A - LC\right)e \tag{11.30}$$

The eigenvalues of the matrix $A - LC$ can be made arbitrarily by appropriate choice observer gain L when the pair of $[A,C]$ is observable, i.e., the observability condition holds. Especially, it can be made Hurwitz, so the observer error $e(t) \to 0$ when $t \to \infty$. Due to the separation principle, k and L can be chosen independently without harm to the overall stability of the systems. As a rule of thumb, the poles of the observer $A - LC$ are usually chosen to converge five times faster than the poles of the system $A - BK$. The observability of the system can be checked by:

$$OM = \left[C\, CA\, CA^2\, ...CA^n\right]^T \tag{11.31}$$

The rank of OM should be equal to four for observability of the rotary inverted pendulum.

11.3.8 LUENBERGER OBSERVER

When the position sensor produces limited noise Luenberger observers are most effective. Sensor noise is a major problem in motion-control systems as well as noise

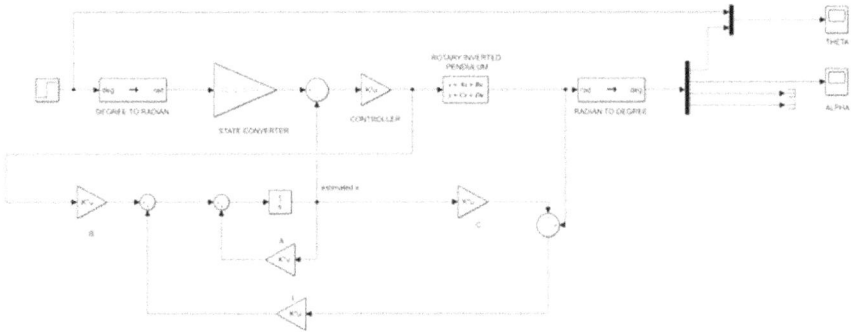

FIGURE 11.2   Simulation of the observer

in servo systems which comes from two major sources: EMI is transmitted to the servo system's control section which is generated by power converters, and resolution limitations in sensors, especially in the feedback sensor. The observer will substantially improve system performance because of the availability of high-resolution feedback sensors [14]. The Luenberger observer is also referred to as a state observer or simply an observer on occasion.

11.3.9   SIMULATION OF OBSERVER

Figure 11.2 shows the observer model of a system with matrix L and estimated state $\hat{x}$. Here the output of a system is taken to the observer. The observer gain for L is calculated from MATLAB workspace through the command $place(A,C,[P_{11}P_{22}P_{33}P_{44}])$. The poles $\left(\left[P_{11}P_{22}P_{33}P_{44}\right]\right)$ are just five times more than the system poles. The desired poles are discussed and the observer only deals with better performances among the desired poles of the controller.

11.4   SIMULATION RESULTS

Simulations have been done using MATLAB: Simulink. The results of full-state feedback controller and state observer are obtained.

11.4.1   FULL-STATE FEEDBACK CONTROLLER RESPONSE

The rotary inverted pendulum was subjected to different controller gain. The gain is changed when the damping ratio (δ) is changed. The arm of the rotary inverted pendulum is applied with 10 degrees of a step input for 5 seconds. Figure 11.3 shows the responses from the controller with a different damping ratio (δ) for the arm angular position. As the damping ratio increases, the overshoot of the arm angular position diminished which is clear from the result.

The main objective is to settle the pendulum's angular position at 0 degrees. Figure 11.4 shows that the pendulum angular position response from the controller.

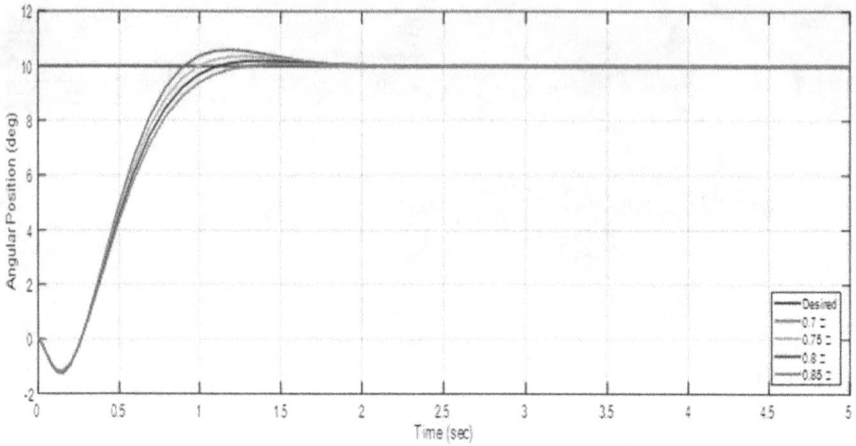

FIGURE 11.3 Simulation of results of the arm angular position θ without the observer

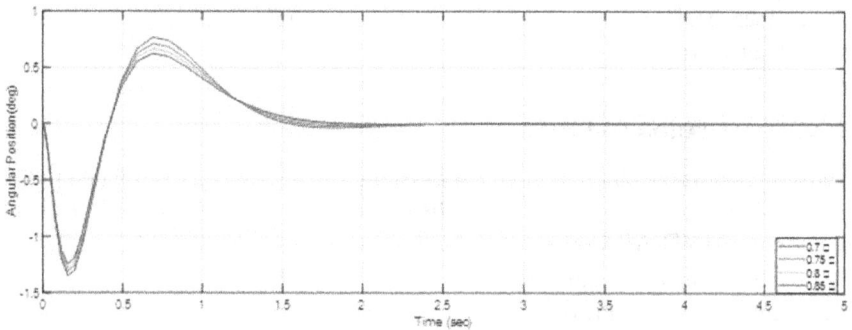

FIGURE 11.4 Simulation of results of the pendulum angular position α without the observer

The damping ratio $\delta = 0.8$ and $\delta = 0.85$ have given better responses in both the arm angular position and pendulum angular position. The fastest response, minimum overshoot, and steady-state error are seen. The controller is given the best system to hold the pendulum at its upright position, minimum overshoot and minimum undershoot to the damping ratio $\delta = 0.85$. From Figure 11.4, it is observed that using a full-state feedback controller pendulum the angle becomes zero within 2 seconds and maintains that position when stable.

11.4.2 State Observer Response

From Figure 11.4, the better response is given for the damping ratio $\delta = 0.85$. For this damping ratio, the observer is designed. Figure 11.5 shows the response of the arm's angular position from the observer. The settling time is lesser than the controller output with the initial condition as zero. Figure 11.6 shows the response of the

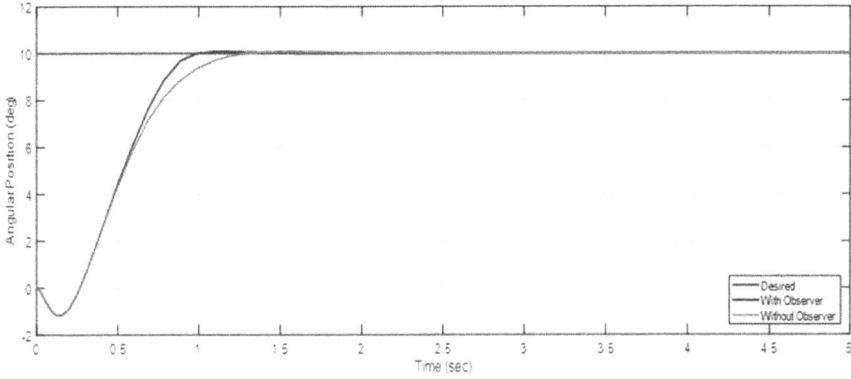

FIGURE 11.5   Simulation results of the arm angular position θ from the observer

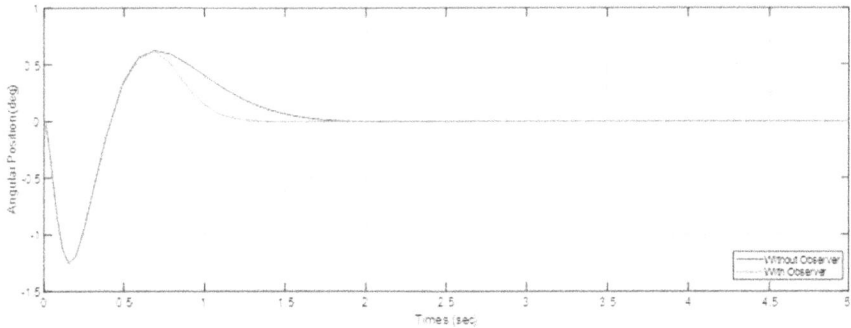

FIGURE 11.6   Simulation results of the pendulum angular position α from the observer

pendulum's angular position from the observer. From the result, it is clear that the observer not only estimates the state but also improves the responses. In the arm angular position, it is like a critical damping. In the pendulum angular position, the pendulum is stable (angle becomes zero) within 1.25 seconds comparatively less than the controller output.

11.5   CONCLUSION

The rotary inverted pendulum is prominently deployed in solving various engineering problems in real-time systems. There are many techniques to stabilize the RIP system, based on robustness and complexity this chapter has proposed a PID controller along with an observer with a set of specifications. The controller is designed with a settling time of 5 seconds, 10-degree step input, $\delta = 0.85$ of damping ratio with $\omega_n = 4$ rad/s. owing to these specifications the pendulum is held at an upright position. Enhancement of system response is accomplished by using an observer. A rigorous analysis has also been given to show the reason why the Luenberger can effectively hold the pendulum at an upright position. This implies that the observer not only reduces the sensor requirements but also improves the system response.

REFERENCES

1. Z. Ding, Z. Li. A Cascade Fuzzy Control System for Inverted Pendulum Based on Mamdani-Sugeno Type. *2014 9th IEEE Conference on Industrial Electronics and Applications (ICIEA)*, 792–797, 2014.
2. K. Ogata. *Modern Control Engineering*, 5th edition, Pearson, 2010.
3. Mehmet Öksüz, Mehmet Burak Önal, Recep Halicioğlu, Lale Canan Dülger. Alternative Controller Design for Rotary Inverted Pendulum. *Technical Journal*, 12(3), 139–145, March 2018.
4. Gou Yun Tao, Du Gang, Wang Jing, Zhang Dongxia. The Design and Realization of a Rotary Inverted Pendulum Based on STM32. *International Conference on Identification, Information, and Knowledge in the Internet of Things (IIKI)*, Beijing, China, 2015.
5. Velchuri Srisha, Dr. Anjali S. Junghare. A Comparative Study of Controllers for Stabilizing a Rotary Inverted Pendulum. *International Journal of Chaos*, 3, 1–13, 2014.
6. Wen-Hua Chen, Jun Yang, Lei Guo, Shihua Li. Disturbance – Observer – Based Control and Related Methods – An Overview. *IEEE Transactions on Industrial Electronics*, 63(2), 1083–1095, February 2016.
7. Abhishek Kathpal, Ashish Singla. SimMechanics TM Based Modeling, Simulation and Real-Time Control of Rotary Inverted Pendulum. *International Conference on Intelligent Systems and Control*, October 2017.
8. Md. Akhtaruzzaman, A. A. Shafie. Control of a Rotary Inverted Pendulum Using Various Methods, Comparative Assessment and Result Analysis. *International Conference on Mechatronics and Automation*, August 4–7, Xi'an, China, 2010.
9. Francesco Ferrante, Frédéric Gouaisbaut, Sophie Tarbouriech. Observer-Based Control for Linear Systems with Quantized Output. *2014 European Controlled Conference (ECC)*, June 24–27, Strasbourg, France, 2014.
10. J. Vonkomer, I. B́elai, M. Huba. Disturbance Observer Control for AC Speed Servo with Improved Noise Attenuation. *Conference: IFAC 2017, World Congress*, Toulouse, France, 2017.
11. Chen Xi-Song, Li Juan, Yang Jun, Wang Jia-Qu. Disturbance Observer Enhanced PID Decoupling Control for Multi-variable Processes. *Proceedings of the 32nd Chinese Control Conference*, July 26–28, China, 2013.
12. N. D. Quyen, N. V. Thuyen, Q. H. Nguyen, D. H. Nguyen. Rotary Inverted Pendulum and Control of Rotary Inverted Pendulum by Artificial Neural Network. *Proceedings of the National Conference on Theoretical Physics*, 37, 243–249, 2012.
13. https://en.wikipedia.org/wiki/State_observer.
14. George Ellis. *Observer in Control System: A Practical Guide*, 1st edition, 213–215, October 16, 2002.

12 Stock Market Prediction Using Machine Learning and Deep Learning Algorithms

K. R. Baskaran and B. Kaviya

CONTENTS

12.1 INTRODUCTION

The economy of the country generally depends upon the stock exchange of that country. The stock market plays a vital role in the economic growth of developing countries such as India. If the stock market rises, then the country's economy also rises. Humans are very curious about their future. Forecasting refers to an approach to predicting what is likely to occur in the future by observing what had happened earlier in the past and what is occurring currently [1]. Due to the changes over time stock market prediction is not an easy task. The first two methods used to forecast stock prices were fundamental analysis and technical analysis [2]. The quantitative and qualitative techniques used to enhance productivity and business gain are referred to as data analytics.

The huge volume of data generated by stock markets forced researchers to predict stock prices in advance. Various Machine Learning algorithms are proposed to predict stock prices such as support vector machine (SVM), boosted decision tree, logistic regression, ARIMA, random forest, linear regression and least-squares support vector machine (LS-SVM). Deep learning consists of artificial neural networks and hybrid fuzzy neural networks to predict stock prices. For the prediction of stock price sentiment analysis has also been considered.

DOI: 10.1201/9781003307716-12

12.2 LITERATURE SURVEY

A. *Stock market prediction using Machine Learning and deep learning techniques*

Machine Learning has the capacity to learn from past experience or historical data. Deep learning is a type of Machine Learning in which a large volume of training data and an artificial neural network with multiple layers are used to achieve the Machine Learning tasks. Only the successful prediction of a stock's future price can yield significant profit.

1) *ARIMA model*

Sheikh Mohammad Idree [1] discussed the ARIMA (Auto Regressive Moving Average) model for the prediction of stock market movement. The ARIMA model is one of the Machine Learning algorithms for performing time series forecasting. Quantitative and qualitative are the forecasting techniques used for data collection. Quantitative forecasting refers to already available data (i.e. historical data) and qualitative forecasting refers to the data collected from the opinions and judgments of experts. The raw oil dataset is considered for prediction. The two indicators used in the stock market are Nifty and Sensex. Continuous and discrete are the two types of time series analysis. A box plot helps in analyzing the stock price variations taking place during successive months and during successive years. The forecasting process uses the auto.arima() function to choose the best parameter. As a result, the chapter [1] suggests that the ARIMA model is good enough for calculating time series data and it can be used in real-world problems.

2) *Support vector machine (SVM)*

The support vector machine algorithm is used in time series stock market prediction. For regression and classification problems, the support vector machine is used which is a supervised Machine Learning algorithm. SVM is mainly used for the minimization of over-fitting problems [3–8]. In [5, 7], historical finance dataset was considered for prediction analysis. To access the effectiveness, six sets of features are designed [7]. The six features are designed as follows:

a) *Price only*

Only the historical prices were taken into account. These prices are used to investigate whether there are patterns in the history of a stock or not.

b) *Human sentiments*

Instead of using all the messages, the SVM method uses only the annotated message and discards other messages.

c) *Sentiment classification*

Built a model which extracts sentiments from the annotated messages. For this process, stop words were removed and lemmatized using CoreNLP.

d) *LDA-based method*

The hidden topic in each message is considered in this model. The documents are represented as random mixtures over latent topics, where each topic is characterized by a distribution over words, thus LDA is a simple topic model to discover the hidden topics.

e) *JST-based method*

The opinion is always expressed as a topic or aspect. Two kinds of models were used to extract the pair of topic sentiments. One of the models is a JST-based method, which is used to extract topics and sentiments simultaneously.

f) *Aspect-based method*

As in the previous models, the mixture of hidden topics and sentiments was considered, whereas in this model the mixtures are not hidden. The support vector machine shows good results in the prediction of stock prices [5, 7], whereas in [3, 4, 6, 8] the SVM does not produce the best performance when compared to other algorithms.

3) *Decision tree*

A decision tree is a supervised Machine Learning algorithm. A decision tree is a hierarchical decomposition of data in which a condition on the attribute value is used to divide the data space hierarchically. Each leaf node is associated with a classification value and each non-leaf node is associated with an attribute feature. Boosting in a decision tree is an iterative technique to adjust the weight of an observation which is based on the last classification such a boosted decision tree was implemented in [3]. In [3] oil, bank and mining, dataset is considered for two models, which are a daily prediction model and a monthly prediction model. A boosted decision tree, logistic regression and support vector machine are used in [3]. By considering a daily prediction model, up to 70% accuracy is determined. A monthly prediction model tries to check whether there is any similarity in attribute values for any two months and proves that there is a similarity in attribute values for various attributes for various months that are not directly related.

In [8], financial datasets of 39 countries were considered. Six algorithms were implemented. The algorithms are logistic regression, decision tree, random forest, extreme gradient boosting, support vector machine, and neural networks. Out of these algorithms, a decision tree is not providing greater accuracy. But in [3], a decision tree shows high accuracy when compared with logistic regression and SVM.

4) *Logistic regression*

Logistic regression is a predictive analysis technique which is based on the concept of probability. It is implemented in [3, 8]. The relationship between the dependent variable and one or more independent variable can be measured with the help of logistic regression by estimating the probabilities using the sigmoid/logistic function. If the response variable takes a binary value (yes or no), then logistic regression is preferred. But as a result [3, 8], logistic regression does not give much accuracy when compared with other algorithms.

5) *Linear regression*

The linear relationship between the independent variable X and dependent variable Y can be represented by linear regression. The case of one explanatory variable is termed a simple linear regression. For more than one explanatory variable, the method is termed multiple linear regression. Linear regression is useful when the dependent variable is binary and it is commonly used in the field of investment for the prediction or classification of companies. The study [6] was implemented using linear regression, random forest, and SVM. Twenty-two years of financial data were considered to design directed and undirected volatility networks of the global stock market based on the correlation and connectedness of national stock indices. Linear regression is compared with random forest and SVM. The result for each prediction year shows that linear regression performs high returns only for short-term forecasting periods, not for long-term periods.

6) *Random forest (RF)*

Random forest belongs to the category of ensemble algorithm. The RF is composed of multiple decision trees, which are randomly selected variables. The combined prediction trees are expected to increase the accuracy and stability of the model performance as compared with those of a single classification model. In the case of RF, the larger the size of the forest (the number of trees), the more the convergence of the generalization error to a specific value and thus, over-fitting can be avoided. RF uses randomly extracted data from the total training dataset and is not significantly affected by noise or outliers. The study [6] used the random forest to avoid over-fitting. In the study [9], stock selection using the Chinese stock market along with stock selection strategies, such as multi-factor and momentum space strategies, were considered. The RF model was implemented on a daily basis dataset which was collected from the Wind financial database that compares the multi-factor, momentum space strategies. It concluded that the momentum feature space is more significant than the multi-factor.

7) *Least-squares support vector machine*

The particle swarm optimization (PSO) algorithm and least-squares support vector machine (LS-SVM) algorithm were integrated in [2] and it is called LS-SVM-PSO.

The PSO algorithm optimizes LS-SVM to predict the daily stock prices based on historical data and technical indicators. The PSO algorithm is used to avoid the local minima problem. The over-fitting problem is also avoided to improve prediction accuracy. The projected model has been evaluated using 13 benchmark financial datasets. It is compared with ANN using the Levenberg–Marquardt (LM) technique. The performance of the LS-SVM-PSO is better than LS-SVM or ANN-BP. LS-SVM-PSO attains a lesser error value over LS-SVM. The ANN-BP algorithm is worse when compared with LS-SVM and LS-SVM-PSO.

8) *Extreme gradient boosting (XGBOOST)*

XGBoost is one of the implementations of gradient boosting to control over-fitting for better performance. The XGBOOST algorithm is a Machine Learning technique for regression and classification issues. XGBOOST is one of the implementation algorithms used in [8] which offers increased efficiency, accuracy, and scalability. It provides better performance in forecasting stock market crisis events.

9) *Artificial neural network*

An artificial neural network is a computational model based on the structure and function of biological neural networks. In the study [2] ANN has been used for stock market prediction for 13 financial benchmarks. ANN suffers from the over-fitting problem and it also does not provide better accuracy when compared to other algorithms.

10) *Deep learning using MXNet*

In the study [8], deep learning is implemented to address the problem of global crisis event forecasting. Multi-layer perceptron is built using MXNet package of R. Financial data of 39 countries were considered in forecasting stock market crisis events. A deep neural network built using MXNet performed best when compared to other networks.

B. *Stock market prediction using sentiment analysis*

Sentiment analysis aims to determine the perspective of a speaker or author with relevance to some topic. It is the process of identifying opinions expressed in a text to determine whether the statement is positive, negative, or neutral. In [10] sentiment analysis was used in stock market prediction. The Machine Learning approach and lexicon-based approach were the two methodologies that are considered. Unsupervised and supervised learning algorithms are the categories of the Machine Learning approach. The lexicon-based approach consists of a dictionary and a corpus approach. The dictionary approach uses a predefined dictionary of words where each word is associated with a specific sentiment polarity strength, whereas the corpus approach is based on a seeding list of opinion words and then finding another opinion word that has a similar context. In this system, sentiment analysis was demonstrated for the stock market by fetching Sensex and Nifty live server data.

12.3   COMPARISON OF EXISTING SYSTEM

In Table 12.1, "stock market predictions using Machine Learning, deep learning and sentiment analysis " which are considered in the survey are summarized.

12.4   PROPOSED WORK

The main aim of the stock trend prediction is to foresee the stock value(s) in the near future. The depreciation of the Indian rupee in recent times has led to the critical need for stock market prediction so as to safeguard the interest of investors. It is planned to predict stock prices for six products (oil, gas, non-sustainable gas, diesel, brent, and gold). All the products used in the work are commodity-based products and are taken from Datahub and Kaggle sites. Each dataset consists of the actual value of the items for the various months of a year. The objective is to find out which of the three algorithms viz., support vector regression, linear regression, and logistic regression gives a better prediction of stock value. All three algorithms that are to be used in the proposed work are of the regression type, because there is only one feature (i.e. year) and many labels (i.e. product price). Using the Python modules all the datasets are combined into a single data frame. The data frames are split into testing and training data and then three algorithms will be used to predict the stock price.

In the chosen dataset, the past stock price for the chosen items is available. 70% of the total dataset is used for training the three algorithms. 30% of the chosen dataset is used for testing. From the testing, it can be shown which of the three algorithms has predicted the stock price of the items more accurately. The best of the above three algorithms can be used for future prediction of the stock values for the chosen items namely (oil, gas, brent, non-sustainable gas, diesel, and gold).

  A. *Support vector regression*

Support vector regression (SVR) is characterized by the use of kernels, sparse solution, and VC control of the margin and the number of support vectors. Although less popular than SVM, SVR has been proven to be an effective tool in real-value function estimation. It uses a support vector machine (SVM, a classification algorithm) algorithm to predict a continuous variable.

  B. *Linear regression*

Linear regression analysis is used to predict the value of a variable based on the value of another variable. The variable to be predicted is called the dependent variable. The variable to predict the other variable's value is called the independent variable. This form of analysis estimates the coefficients of the linear equation, involving one or more independent variables that best predict the value of the dependent variable.

TABLE 12.1

Comparative Study of Various Algorithms for Stock Market Prediction

Ref.no	Algorithm/ Methodology	Dataset	Result	Future Work
Sheikh Mohammad Idrees [1]	ARIMA model	Raw oil (5 years: Jan 2012–Dec 2016)	ARIMA model is good enough for calculating time series data.	It can be used in real-world problems such as that of the health sector, education, finance, and other practical domains for prediction.
Aparna Nayak [3]	1. Boosted decision tree 2. support vector machine 3. Logistic regression	Oil, bank, mining sector, historical, and social media data	Decision-boosted tree performs better than support vector machine and logistic regression.	By considering sentiments, monthly predictions can be made more accurate.
V Kranthi Sai Reddy [4]	Support vector machine	Finance dataset	The SVM algorithm works on the large dataset value which is collected from different global financial markets. The over-fitting problem is avoided.	—
Ping-Feng Pai [5]	1. ARIMA Model 2. SVM Hybrid model	Company-based dataset (Oct 2002–Dec 2002)	Different prediction models can complement each other on exact data sets, and proposed a hybrid model of the ARIMA and the SVM.	—
Tae Kyun Lee [6]	1. Linear regression 2. RF 3. SVM	22-year dataset from Jan 1995–Dec 2016	Mid-term investments with short-term volatility showed better performance than other strategies in terms of stock market prediction.	Used only a connectedness measure but we must develop other network indicators for market capture. Three Machine Learning techniques are not sufficient, deep learning should be considered.

(*Continued*)

TABLE 12.1 (CONTINUED)
Comparative Study of Various Algorithms for Stock Market Prediction

Ref.no	Algorithm/ Methodology	Dataset	Result	Future Work
Thien Hai Nguyen [7]	Support vector machine	1. Historical dataset (Yahoo 18 stocks). 2. Message board dataset (one year from 23 July 2012–19 July 2013)	This research used SVM for Stock Market prediction. 60% accuracy is achieved.	Only historical prices and sentiments from social media are considered, in the future attempts to find and integrate more features will be made, which can affect the stock prices to develop a more accurate stock prediction model.
Sotirios P.Chatzi [8]	1. Logistic regression 2. Random forest 3. SVM 4. Neural network 5. Extreme gradient boosting 6. Deep learning	Finance dataset	MXNet (deep learning) provides the best performance followed by the XGBoost methodology.	—
Zheng Tan [9]	Random forest (RF)	Financial dataset	Developed a stock market strategy based on RF model and implemented in the Chinese stock market. The momentum feature is more significant for performance evaluation.	Further work can be dedicated to more valuable feature spaces and to the development of novel Machine Learning algorithms for stock selection strategies on a daily frequency
Osman Hegazy [2]	LS-SVM is compared with ANN	13 financial benchmark datasets	LS-SVM-PSO achieves the Lowest error rate over LS-SVM. ANN-BP algorithm is not accurate.	—
Aditya Bhardwaj [10]	Sentiment analysis	Nifty and Sensex (TimesofIndia.com)	Demonstrated sentiment analysis for the stock market by fetching Sensex and Nifty live server data.	Future work is to run the Python script code with more advanced functions.

Linear regression fits a straight line or surface that minimizes the discrepancies between predicted and actual output values. There are simple linear regression calculators that use a "least-squares" method to discover the best-fit line for a set of paired data. Then estimate the value of X (dependent variable) from Y (independent variable).

C. *Logistic regression*

Logistic regression is a predictive analysis. Logistic regression is used to describe data and to explain the relationship between one dependent binary variable and one or more nominal, ordinal, interval, or ratio-level independent variables. Logistic regression measures the relationship between the categorical dependent variable and one or more independent variables by estimating probabilities using a logistic/sigmoid function. Here the output is binary or in the form of 0/1 or −1/1.

D. *Diagrammatic representation*

Phase 1:

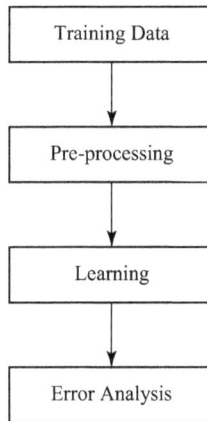

FIGURE 12.1   Phase 1 of prediction

Phase 2:

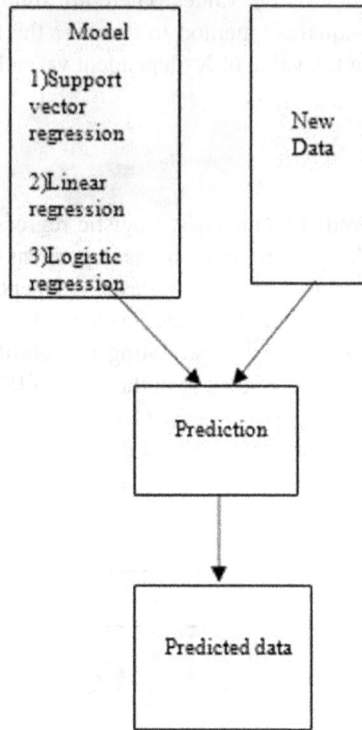

FIGURE 12.2   Phase 2 of prediction

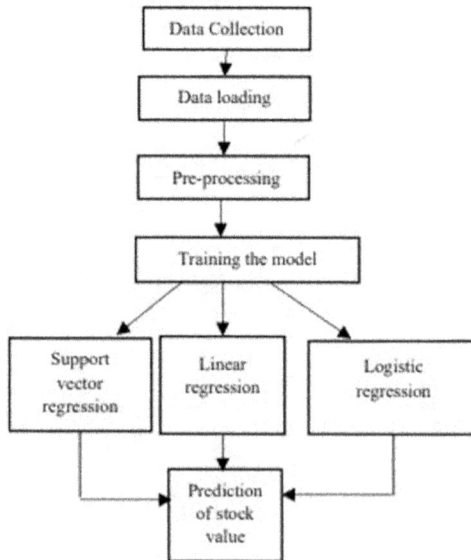

FIGURE 12.3   Work flow

12.5 CONCLUSION

In the literature, different Machine Learning algorithms were chosen by different authors for predicting the stock value of financial and commodity datasets. No common Machine Learning algorithms were chosen by all the authors. In the proposed work, it is planned to consider Machine Learning algorithms such as support vector regression, linear regression, and logistic regression, for stock value prediction and to select the best algorithm for the chosen dataset. In the future, the same three Machine Learning algorithms can be considered for various other datasets and to predict the best Machine Learning algorithm for the prediction of stock value.

REFERENCES

1. Idrees, S. M., Alam, M. A., & Agarwal, P. (2019). 'A prediction approach for stock market volatility based on time series data'. *IEEE Access*, *7*, 17287–17298.
2. Tan, Z., Yan, Z., & Zhu, G. (2019). 'Stock selection with random forest: An exploitation of excess return in the Chinese stock market'. *Heliyon*, *5*(8), 1–10.
3. Nayak, A., Pai, M. M., & Pai, R. M. (2016). 'Prediction models for Indian stock market'. *Procedia Computer Science*, *89*, 441–449.
4. Reddy, V. K. S. (2018). 'Stock market prediction using machine learning'. *International Research Journal of Engineering and Technology (IRJET)*, *5*(10), 1033–1035.
5. Pai, P. F., & Lin, C. S. (2005). 'A hybrid Arima and support vector machines model in stock price forecasting'. *Omega*, *33*(6), 497–505.
6. Lee, T. K., Cho, J. H., Kwon, D. S., & Sohn, S. Y. (2019). 'Global stock market investment strategies based on financial network indicators using machine learning techniques'. *Expert Systems with Applications*, *117*, 228–242.
7. Nguyen, T. H., Shirai, K., & Velcin, J. (2015). 'Sentiment analysis on social media for stock movement prediction'. *Expert Systems with Applications*, *42*(24), 9603–9611.
8. Chatzis, S. P., Siakoulis, V., Petropoulos, A., Stavroulakis, E., & Vlachogiannakis, N. (2018). 'Forecasting stock market crisis events using deep and statistical machine learning techniques'. *Expert Systems with Applications*, *112*, 353–371.
9. Tan, Z., Yan, Z., & Zhu, G. (2019). 'Stock selection with random forest: An exploitation of excess return in the Chinese stock market'. *Heliyon*, *5*(8), August 2019, 1-10
10. Bhardwaj, A., Narayan, Y., & Dutta, M. (2015). 'Sentiment analysis for Indian stock market prediction using sensex and nifty'. *Procedia Computer Science*, *70*, 85–91.

12.5 CONCLUSION

In the literature, 20 different Machine Learning algorithms were chosen by different authors for predicting the stock value of financial and automobile datasets. Not common Machine Learning algorithms were chosen by all the authors in the proposed work. It is planned to consider Machine Learning algorithms such as Support vector regression, linear regression, and logistic regression for stock value prediction and to select the best algorithm for the chosen dataset. The future the same three Machine Learning algorithms can be considered for various input changes and to find out the best Machine Learning algorithm for the prediction of stock value.

REFERENCES

13 Service Line Fault Detector and Distance Calculator for Smart Cities

Ramasamy Dhivya Praba, Kanakaraj Kavitha,
Kandhasamy Thilagavathi, Kumarasway Jasmine,
and S. N. Shivappriya

CONTENTS

13.1 INTRODUCTION

Electrical power system lines are mainly used to transmit electrical energy from the generation unit to the customers. Electrical cables run underground instead of overhead lines in urban areas. Overhead lines are not preferred for electrical energy transmission due to their instability during climatic conditions. Underground cables are resistant and they are not affected by weather conditions. Due to its submerged nature, it is very difficult to find the location of the fault when it occurs. This makes the entire power system damaged. Due to marvelous research improvements in the field of digital systems, fault identification becomes simple if we use digital techniques. In urban areas power can be supplied using an underground cable system. Once the fault is identified, the repair process becomes difficult as the fault location is unknown.

The main aim of the proposed work is to locate the fault from the base station. It is difficult to know the exact location of the cable fault when the cable is laying underground. With the help of IoT technology, fault location is serially communicated to the server. This is an unsolved problem till now. When we predict the fault manually it affects the efficiency of the under-laying cable. Many methods to resolve this problem have been proposed. The challenge is the lack of methods to detect the cable fault and access methods whenever required because it is underground. So here we propose a solution to identify the fault location using IoT through which it can be communicated to the central server.

13.2   LITERATURE REVIEW

Carrying out a literature review early, at the beginning of the research project is essential to understand faults in underground cable lines, as this will supply the researcher with much-needed additional information on the methodologies and technologies available and used by other research complement around the world.

The paper [1] says cable faults are damage to cables which affects the resistance in the cable. If allowed to persist, this can lead to a voltage breakdown. To locate a fault in the cable, the cable must first be tested for faults. This prototype uses the simple concept of Ohm's law. The current would vary depending on the length of the fault of the cable. This prototype is assembled with a set of resistors representing cable length in kilometers and fault creation is made by a set of switches at every known kilometer (km) to crosscheck the accuracy of the same. The fault occurring at what distance and which phase is displayed on a 16×2 LCD interfaced with the microcontroller.

The program is burned into the ROM of a microcontroller. The power supply consists of a step-down transformer 230/12V, which steps down the voltage to 12V AC. This is converted to DC using a Bridge rectifier. The ripples are removed using a capacitive filter and it is then regulated to +5V using a voltage regulator 7805 which is required for the operation of the microcontroller and other components.

The paper [2] proposes a fault location model for underground power cables using a microcontroller. The aim of this project is to determine the distance of the underground cable fault from the base station in kilometers. This project uses the simple concept of Ohm's law. When any fault, such as a short-circuit, occurs voltage drop will vary depending on the length of the fault in the cable since the current varies. A set of resistors are therefore used to represent the cable and a DC voltage is fed at one end and the fault is detected by detecting the change in voltage using an analog-to-voltage converter and a microcontroller is used to make the necessary calculations so that the fault distance is displayed on the LCD.

The aim of project [3] is to determine the underground cable fault. This project uses the simple concept of circuit theory. When any fault, such as a short circuit, occurs voltage drop will vary depending on the length of the fault in the cable, since the current varies circuit theory is used to calculate the variance. The signal conditioner manipulates the change in voltage and a microcontroller is used to make the necessary calculations so that the fault distance is displayed by IoT devices.

13.2.1 TYPES OF FAULTS

The fault in a cable can be of two types:

- Open-circuit fault
- Short-circuit fault

Open-circuit fault:
   Any damage/break in the conductor of a cable is known as an open-circuit fault.

Short-circuit fault:
   It is defined as when two conductors of a multi-core cable are available in electrical contact with each other due to insulation failure. It is further divided into two types:

- Symmetrical fault: In this type, all three phases are short-circuited
- Unsymmetrical fault: In this fault, the magnitude of the current is not equally displaced across 120 degrees

13.3   EXISTING SYSTEM

13.3.1 TRACER METHOD

This can be done by checking the entire cable by walking till the fault is identified. Either an electro-magnetic or audio frequency signal is used for fault detection. It predicts the accurate location of the fault. It is further divided into:

1. Tracing current
2. Sheath coil

13.3.2 TERMINAL METHOD

Without tracing it can detect the fault area, but not the exact location of the fault. It is further divided into:

1. Murray-loop method
2. Impulse-current method

Disadvantages of the existing system:

1. Underground cables have a higher initial cost and insulation problems at high voltages
2. Buried nature of the underground cable, it becomes very difficult to locate and repair the fault

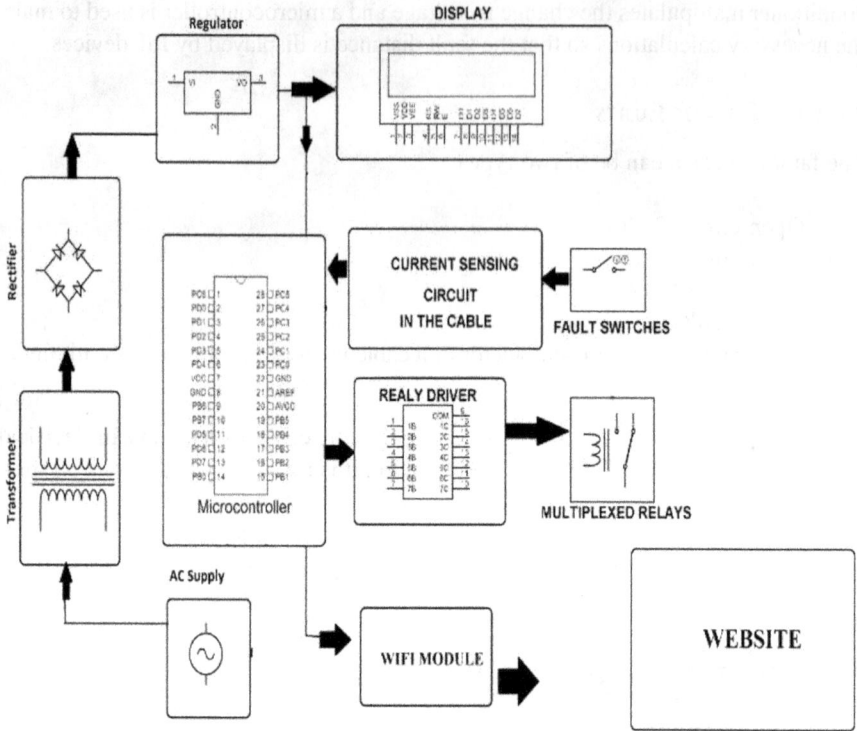

FIGURE 13.1　Block diagram of the proposed system

13.4　MATERIAL AND METHODS

The proposed system uses IoT technology to locate the fault in any underground service line (Figure 13.1, Figure 13.2). Whenever a fault occurs, the microcontroller senses it and the location of the fault is displayed on the LCD screen placed in the control room and the same is displayed on the webpage using IoT technology.

The entire system is connected through various connectors. The relays are connected to a Raspberry Pi Zero microcontroller which acts as an interface. The Raspberry Pi requires a power supply to make the sensors start working. Raspberry Pi provides the power supply to the LCD as well as the relay. The microcontroller is connected to the computer using a USB cable.

The Ubidots framework implements WebSocket using Action Cable. Once the Action Cable is established, there will be a full-duplex communication channel between the web server and the Raspberry Pi. This channel exists until it is manually stopped, hence real-time data reading can be achieved. When a fault occurs (single fault Figure 13.1), the Raspberry Pi detects it, and the details are displayed on the LCD. Green, and red LEDs are used on the webpage to display the fault condition. When multiple faults, shown in Figure 13.2, occur the details are displayed one by

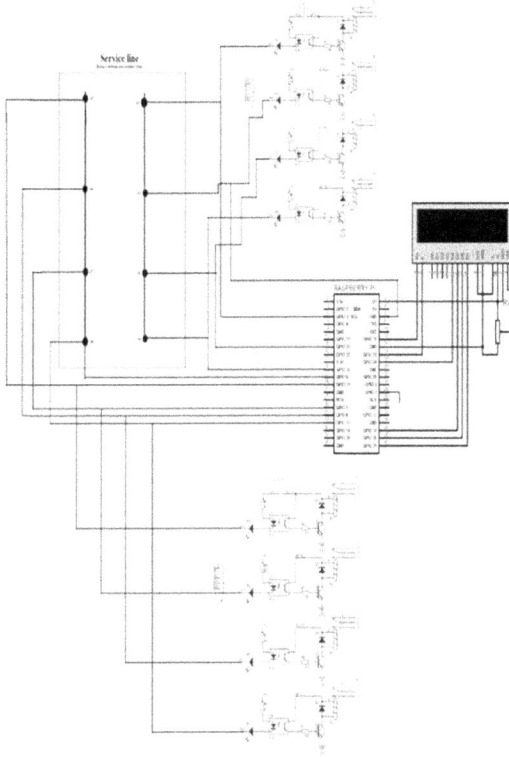

FIGURE 13.2   Circuit diagram of the proposed system

one on the LCD and in IoT all the faults are displayed at the same time. Only when the fault is rectified, the Green LED will blink on IoT and the relay will be powered.

13.5   SOFTWARE USED

13.5.1   ADVANCED IP SCANNER

To use and runs as a portable edition.

13.5.2   MOBAXTERM

The eventual toolbox used for remote computing is MobaXterm. It is very popular for single-window applications. It is user-friendly (Figure 13.3). It is famous and supports many functions for different users such as software engineers, users can handle the environment completely and simply while sitting in a remote place. It supports many remote-controlled network tools to Windows desktops equipped in a single portable .exe file which works out of the box. A few plugins are used to add functions to MobaXterm such as UNIX commands (bash, ls, cat, sed, grep, awk, and rsync).

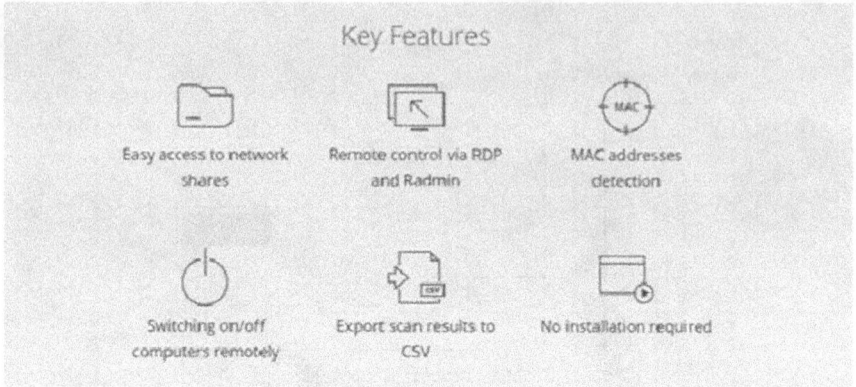

FIGURE 13.3   IP scanner

13.6   RESULTS AND DISCUSSION

The output obtained from the corresponding relay will help us to determine the fault location in an interconnected system in which the fault location is exposed. The output can be observed in the LCD as well as the web page using IoT technology.

13.6.1   MULT-FAULT (IoT):

In Figure 13.4, it is clearly shown in the browser that there are faults in the second and third sections of Line 1 and the third section of Line 2, hence they are shown in a darker colour (reddish).

FIGURE 13.4   Multi-fault

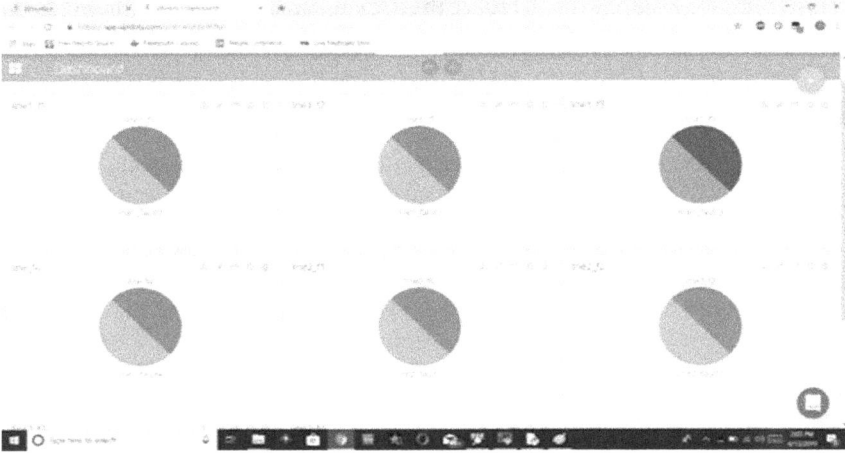

FIGURE 13.5   Single-fault

13.6.2   SINGLE-FAULT (IoT)

In Figure 13.5 it is clearly shown in the browser that there are faults in the third section of Line 1, hence they are shown in LED. The fault is also displayed on the LCD (Figure 13.6).

13.7   CONCLUSION AND FUTURE ENHANCEMENT

The proposed system helps us to locate the fault in an underground service line. The details are uploaded to the IoT platform which can be viewed from anywhere. For underground distribution networks fault location and their state estimation is very challenging. This work may help in some degree to support further analytical and practical studies in the fields of fault location and state calculation for real underground distribution systems by combining the proposed method and the existing tracer method, a hybrid model can be developed which can help in detecting the exact point of the fault. A hybrid model of such experiments can be used in the

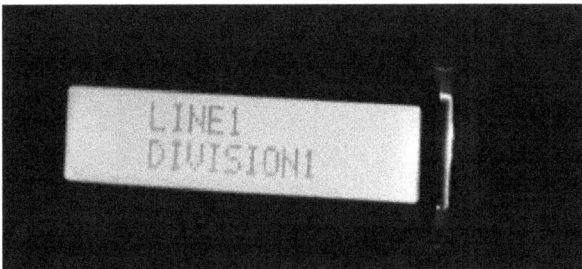

FIGURE 13.6   Fault display LCD

extensive service zone and it will reduce the tracking time and improve the efficiency of the overall system.

REFERENCES

1. Q. Shi, U. Troeltzsch, O. Kanoun. Detection and localization of cable faults by time and frequency domain measurements. *Conference Systems and Signals and Devices, 7th International Conference, Amman*, pp. 1–6, 2010.
2. B. Clegg, *Underground Cable Fault Location*. New York: McGraw-Hill, 1993.
3. M.-S. Choi, D.-S. Lee, X. Yang. A line to ground fault location algorithm for underground cable system. *The Korean Institute of Electrical Engineers*, pp. 267–273, June 2005.

14 Heart Disease Classification Using Multi-Layer Perceptron (MLP) Neural Network

S. N. Shivappriya, Rajaguru Harikumar,
Krishnamoorthi Maheswari,
and Ramasamy Dhivya Praba

CONTENTS

14.1 INTRODUCTION

Ventricular arrhythmias are considered irregular heart rhythms that occur in the bottom chambers of the heart (ventricles). These forms of arrhythmias cause the heart to pump too quickly, which prevents the supply of oxygen-rich blood to the brain and body and can result in cardiac arrest or cardiomyopathy [1]. The PhysioNet database [2] consists of ventricular arrhythmias such as: Myocardiac Infraction (MI), Premature Ventricular Contraction (PVC), Ventricular Tachyarrhythmia (VT), Supra Ventricular Tachyarrhythmia (SVT), ST Elevation (ST) and Malignant Ventricular Arrhythmia (MVA).

Lin [3] shows the frequency domain analysis of the Ventricular Late Potential (VLP) which is the major cause of Sudden Cardiac Death (SCD). Hohlfeld et al. [4],

Mahamoodabadi [5] and Shantha Selva Kumari et al. [6] proposed an algorithm based on wavelet transform for denoising and detection of the ECG characteristic points. David Cuesta-Frau et al. [7], Nikolaev et al. [8] and Umamaheswar Reddy et al. [9] developed and evaluated a robust wavelet transform–based ECG denoising system. Shivappriya [10], Shivappriya and Harikumar [11], proposed a Stationary Wavelet Transform (SWT), for the detection of ECG signal fiducial points, where these points are connected to characteristic waves such as the QRS-, P- and T-wave complexes.

The developed algorithm is to be used to save human lives, hence it is preferable to use the supervised learning approach in classifying cardiac arrhythmias with the extracted features. To classify cardiac arrhythmia four different supervised neural network classifiers are used, Probabilistic Neural Network (PNN), Back Propagation Neural Network (BPNN), Extreme Learning Machine (ELM) and Self-adaptive Resource Allocation Network (SRAN) Classifier.

14.2 MATERIALS AND METHODS

The functioning of the heart is analyzed with the ECG signal. The difference in heart functioning is determined by the width and height of the ECG signal. While in the fetching process, the ECG signal interferes with electrode contact noise, power line interference, motion artifacts, channel noise, baseline wander, and electromyogram (EMG) noise. The ECG signal should be denoised using Stationary Wavelet Transform (SWT) [12], then 16 morphological and statistical features are extracted. This chapter uses one lead ECG signal (II) that contains six different arrhythmias together with a normal ECG signal for training. Four different Neural Network classifiers [13] are used such as the Probabilistic Neural Network (PNN), Back Propagation Neural Network (BPNN), Extreme Learning Machine (ELM) and Self-adaptive Resource Allocation Network (SRAN) classifier.

14.3 ECG SIGNAL PROCESSING THROUGH WAVELET TRANSFORM

With this Wavelet Transform (WT) the ECG signal can be represented with time, frequency and amplitude localization. The part of the ECG signal is extracted based on its frequency and amplitude multiresolution characteristics of the P-, QRS- and T-waves [8]. Stationary Wavelet Transform (SWT) [10] is a redundant transform used to identify predominant features of the signal at a precise location with an improved signal-to-interference ratio. The biorthogonal waveform looks similar to the ECG waveform, it is preferred to locate the exact local maxima and minima.

Biorthogonal wavelet system: The signal spans need not be orthogonal it has to be invertible and also non-orthogonal wavelet bases. Signal, "s" represented using biorthogonal wavelet system:

Let $f(t)$ be a signal (function) from which V_j is given by:

$$f(t) = \sum_k s_j(k) 2^{j/2} \varphi(2^j t - k) \tag{14.1}$$

Since $\varphi_{j,k}(t)$ are not orthogonal, to get $s_j(k)$, project $f(t)$ on to the dual basis $\varphi_{j,k}(t)$ i.e.,:

$$s_j(k) = \int f(t)2^{j/2}\tilde{\varphi}(2^j t - k)dt = \int f(t)\tilde{\varphi}_{j,k}(t)dt \tag{14.2}$$

This $f(t)$ is projected onto its dual basis.

WT is a novel mathematical tool for signal analysis and pattern recognition. Input features are extracted using a wavelet transform algorithm [11]; input features which are derived from the 240 ECG signals are shown below:

QRS detection and delineation time; R and P wave amplitudes; P-wave detection and delineation time; R, S and T amplitudes; T-wave detection and delineation time; RR interval; slope of ST interval; mean; variance; skewness; kurtosis; standard deviation; and spectral entropy, which are taken from the six different arrhythmia databases.

14.4 SUPERVISED MACHINE LEARNING APPROACH

The dataset consists of input features where the target class will train the neural network, if the unknown class input features are given as input to the network, then the network will correctly identify the class label. Four different supervised Machine Learning networks are used to validate the classifier performance.

14.4.1 EXTREME LEARNING MACHINE (ELM)

Based on the network topology the ELM is also known as a Single Hidden Layer Feedforward Neural Network (SLFN) [14], extracted features are given to the input layer, hidden layer is used to map the features with its related classes and the output layer predicts the arrhythmia classes. In the forward pass of the classification process, the predicted output has been labeled as $\left|\hat{Y}\right|$. In the backward pass, the error has been calculated from the difference between the target (Y) and the predicted output $\left|\hat{Y}\right|$ With respect to these errors the weight updating takes place randomly between the input and the hidden layer. The gradient-based weight optimization technique has been used to train NNs. The output of the hidden neuron will have the ability to classify the predicted output $\left|\hat{Y}\right|$ with the target (Y) class with the controlling weight between the hidden layer and the output layer.

In Equation (14.3) the input weight is W, the bias of hidden neurons is B and the output weight is O. The output of the ELM network is:

$$\hat{Y}_{ik} = \sum_{j=1}^{N} O_{kj} G_j\left(W, B, X_i\right), k = 1, 2, \ldots, C \tag{14.3}$$

The output of the jth hidden neuron is:

$$G_j = G\left(\sum_{k=1}^{n} W_{jk} x_{lk} + b_j\right), j = 1, 2, \ldots, H \tag{14.4}$$

where *G(.)* is the activation function. If it is a Radial Basis Function (RBF) [15], the output is:

$$G_j = G\left(b_j \parallel X - W \parallel\right), j = 1, 2, \ldots, H \tag{14.5}$$

W is the center and b_j is the width of the neurons in RBF.

The objective of this ELM approach is to show the relationship between the features with their class labels. This decreases training time, has a good generalization capacity and a fast learning capability.

14.4.2   SELF-ADAPTIVE RESOURCE ALLOCATION NETWORK (SRAN) CLASSIFIER

The SRAN classifier follows a sequential learning algorithm with a self-regulated control mechanism built into it [16], it uses both productive and fixed architecture networks. Training samples are used one by one and only once, it computes first the difference in current sample information from the information obtained by the network. The samples are entered into the training or pushed into the stack if the computed difference is too large, else it will be deleted from the training set to avoid over-training.

The network output for the current input feature *(xi)* is:

$$y_j = w_{jk}\emptyset_k\left(x_i\right), i = 1, 2 \ldots .n \tag{14.6}$$

For the input x_i, the activation function of the kth hidden neuron is $\emptyset_k(x_i)$ and the connection weights of the hidden neuron *k* and the *j* output neuron is w_{jk}. The bias term to the jth output neuron is α_{j0}. The Gaussian equation is given by Equation (14.7).

$$\emptyset_k\left(x_i\right) = e^{\frac{\left(\parallel x_i - \mu_k \parallel\right)}{(\sigma_k)^2}} \tag{14.7}$$

SRAN produces a complete network with better generalization and reduces training time and training samples significantly.

14.4.3   BACK PROPAGATION NEURAL NETWORK CLASSIFIER (BPNN)

Introducing more hidden layers into the feedforward neural network will frame the multi-layer neural network structure. Based on the complexity of the problem the number of hidden neurons and hidden layers is increased. A differentiable activation function is needed for the learning algorithm for multi-layer perceptron's, also used as a logistic function (non-linear, monotonic, increasing and differentiable) [17]. In the forward pass, the error value is calculated between the input layer, hidden layer and output layer. The term back propagation means from the output layer to hidden layer then hidden layer to the input layer the weights are optimized with respect to the error (target − actual output) value derived from the forward pass of the neural network [18, 19]. In the training phase optimized weights are developed and testing of the new features with the unknown outputs is verified/validated.

14.4.4 Probabilistic Neural Network (PNN)

It is easy to use and is extremely fast for classifying moderate-sized databases; it is faster than MLP [20]. Classification is done with a four-layer neural network architecture. The extracted features are given to the input layer, the pattern layer estimates the Probability Density Function (PDF) a using multi-dimensional Gaussian kernel and the summation layer combined all the PDF of the patterns which belong to the same class and the output layer shows that categories the arrhythmias based on its probability. The classifier output is insensitive to outliers; the predicted probability score is more precise.

14.5 PERFORMANCE MEASURE

With the following performance measure, the feature extraction and neural network classifiers are analyzed.

$$\text{Miss} = \frac{FN}{TP + FN} \times 100 \tag{14.8}$$

$$\text{Fallout} = \frac{FP}{TN + FP} \times 100 \tag{14.9}$$

$$\text{Specificity} = \frac{TN}{TN + FP} \times 100 \tag{14.10}$$

$$\text{Precision} = \frac{TP}{TP + FP} \times 100 \tag{14.11}$$

$$\text{Accuracy} = \frac{TP + TN}{TP + FN + FP + TN} \times 100 \tag{14.12}$$

$$\text{Error beats } E = FN + FP \tag{14.13}$$

Where True Positive (TP), False Negative (FN), False Positive (FP) and True Negative (TN) are the number of events that are observed, erroneously rejected, erroneously confirmed and correctly rejected. The correct rate is the ratio of correctly classified samples to the classified samples. The error rate is the ratio of wrongly classified samples to the classified samples. The last correct rate is the ratio of the correctly classified samples to the classified samples last error rate.

The last error rate is the ratio of wrongly classified samples to the classified samples. The inconclusive rate is the ratio of non-classified samples to the total number of samples. The classified rate is the ratio of classified samples to the total number of samples. The negative predictive value is the ratio of correctly classified negative samples to negative classified samples. The negative likelihood is the ratio of, 1 − sensitivity, to the specificity. The prevalence is the ratio of true positive samples to total number of samples.

14.6   RESULTS AND DISCUSSION

The primary objective of ECG analysis is the diagnosis of the disorder. In general, the diagnosis of an automatic disease is done by following the limits of the ECG parameters. The reliability of the diagnosis of the disease is therefore totally dependent on the accuracy of the estimates of the ECG parameters. In this chapter, a delineator using four different neural network classifiers is proposed and evaluated on different arrhythmia databases. The proposed algorithm compares two different databases (with 11 and 16 features) to validate the accuracy of the algorithm. The doctor classifies arrhythmia with rhythm and morphology information; an input vector should include properly representing the rhythm and morphology. To extract these, two different feature extraction techniques have been implemented by using the Pan–Tompkins method and wavelet-based delineation algorithm. Table 14.1 shows a performance comparison of these feature extraction techniques [21].

T-Wave Alternations (TWAs) are consistent fluctuations in the T-waves, on every other beat basis. The Low Pass Differentiator (LPD)–based method gives 97.74% of sensitivity and won't give perfect T-wave offset point detection, which is simple in implementation and robust to waveform variations. The sensitivity of the proposed approach is 99.83% which is comparatively higher than the LPD method and DWT (99.77%) method. Figure 14.1–14.4 show different T-wave morphologies of different ventricular arrhythmia Diseases [22]. With the feature extraction method, different statistical and morphological features of the ECG signal are extracted as follows:

TABLE 14.1
Performance Comparison of Feature Extraction Techniques

(For 33 Recorded MITBIH Data)	Pan–Tompkins Method		Wavelet-Based Delineation Method	
	Sensitivity (%)	Positive Predictivity (%)	Sensitivity (%)	Positive Predictivity (%)
QRS Complex Detection	95.76	99.63	93.45	99.97
QRS Complex Delineation	95.73	99.24	93.45	99.7
T-wave Detection	95.74	99.19	93.57	98.73
T-wave Delineation	99.97	99.25	99.83	99.84
P-wave Detection	95.79	98.5	93.43	99.24
P-wave Delineation	99.92	98.54	99.97	99.32

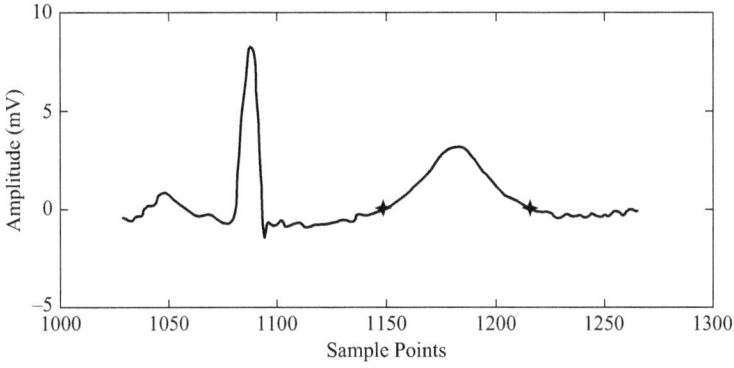

FIGURE 14.1   Positive T Wave

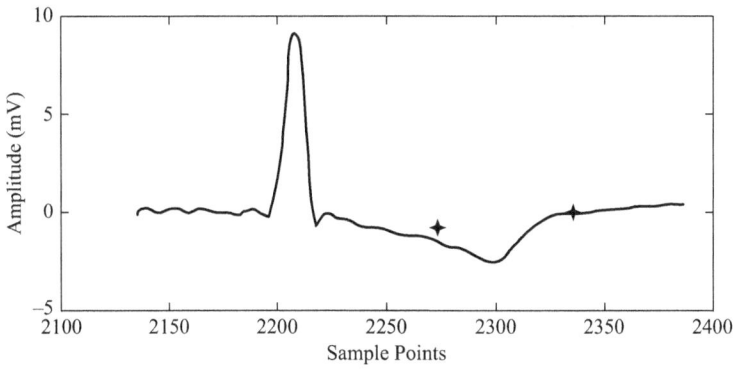

FIGURE 14.2   Negative T Wave

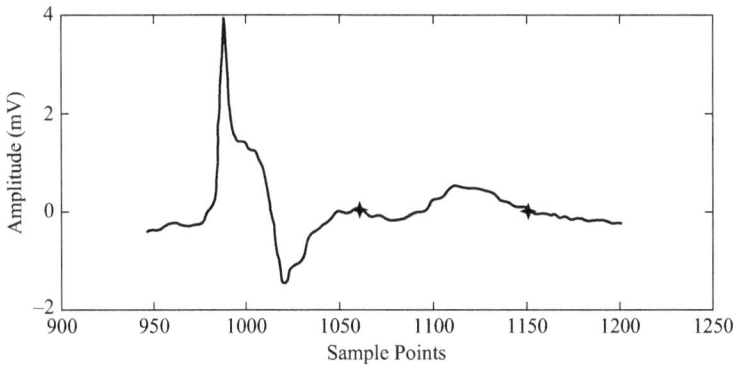

FIGURE 14.3   Biphase(+/− or −/+) T Wave

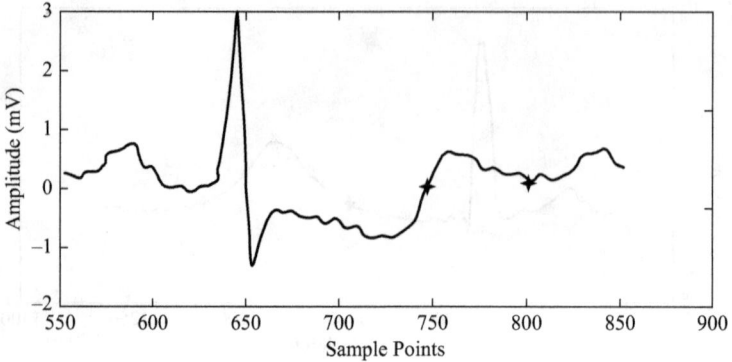

FIGURE 14.4   Ascending T Wave

QRS time, QRS delineation time, R-wave time, R-wave delineation time, P-wave amplitude, P-wave delineation time, R-amplitude, S-amplitude, T-amplitude, T-wave delineation time, RR interval, the slope of ST interval, mean, variance, skewness, kurtosis and standard deviation, spectral entropy.

To the NN classifier, the above features are supplied as inputs. Training, validation and test sets were formed by 240 samples (40 samples from each class) of 11 dimensions (dimension of the extracted feature vectors) (Dataset 1) and 140 samples (20 samples from each class) of 16 dimensions (Dataset 2). Dataset 1 comprises MI, PVC, VT, SVT, ST and LT. Dataset 2 comprises MI, PVC, VT, SVT, ST, MVA and Normal.

The different learning method in Table 14.2 shows that the gradient descent algorithm gives a higher classification rate than other learning methods [23]. This is suitable for dynamic datasets and occupies less memory. Experimental result shows that the feature vector extracted by the WT can characterize ECG patterns, and MLP NN classifier with gradient descent algorithm back propagation algorithm has a stronger ECG recognition effect with 85.8% accuracy. The classification results of the PNN used for classification of different CAs with Dataset 1 and Dataset 2, as given in Table 14.3.

For all the datasets the classification rate and specificity are higher and missed classification rate, error rate and last error rate values are lower. From each class, 40

TABLE 14.2

Performance Comparison of Three Different Learning Methods with BPNN Classifier

Learning Method	Performance Measure	BPNN Classifier
Levenberg–Marquardt Back Propagation	Error rate	0.133
	Correct rate	0.699
Gradient Descent Back Propagation with Momentum and Adaptive Learning	Error rate	0.141
	Correct rate	0.858
Resilient Back Propagation	Error rate	0.135
	Correct rate	0.750

TABLE 14.3

Performance Metrics of PNN Machine Learning Classifier

Performance Measure	Dataset 1	Dataset 2
Misclassification Rate	0.29	0.15
Correct Rate	0.72	0.85
Last Correct Rate	0.72	0.85
Inconclusive Rate	0.83	0.85
Specificity	0.73	0.85
Error Rate	0.28	0.15
Negative Predictive Value	0.13	0.12
Last Error Rate	0.28	0.15
Negative Likelihood	1.39	0.18
Classified Rate	0.17	0.14
Prevalence	0.83	0.86

TABLE 14.4

Comparison of Different Learning Methods of ELM Machine Learning Classifier

Database	Activation Function	Training Accuracy (%)	Testing Accuracy (%)
Dataset 1	Unipolar	100	93.33
	Bipolar	100	85
	Radial basis function	100	91.67
Dataset 2	Unipolar	65.83	68.33
	Bipolar	65	63.33
	Radial basis function	64.16	64.16

signals are chosen for classification, with a total of 240 samples, 168 for training, 36 for validation and 36 samples for testing are selected. Around 1000 iterations are performed with different training, validation and testing percentages [(70, 15, 15), (80, 10, 10), (60, 20, 20)]. Table 14.4 shows various functions most used in neural networks. The performance comparison of four different NN classifiers is shown in Table 14.5 for two different data and feature sets. Among the four classifiers the PNN classifier gives 99.8% accuracy for the 16-feature dataset.

14.7 CONCLUSION

In order to save human life, the accurate detection and classification of acute ventricular arrhythmia has been developed with a higher false alarm suppression rate. Maximum wavelet coefficients were extracted based on the WT-based denoised and delineated ECG signal. Then these extracted morphological and statistical features

TABLE 14.5

Performance Comparison of Neural Network Machine Learning Classifier

	Accuracy (%)	
Classifier	Dataset 1	Dataset 2
Extreme Learning Machine (ELM)	67	71
Back Propagation Neural Network (BPNN)	71.7	85.8
Self-Adaptive Resource Allocation Network (SRAN)	75	76
Probabilistic Neural Network (PNN)	72	99.8

of ECG signals are to be provided as input to the NN Machine Learning classifier. Experimental result shows that the feature vector extracted by the WT can characterize ECG patterns, and the MLP NN classifier with gradient descent algorithm and back Propagation algorithm has a stronger ECG recognition effect with 85.8% accuracy. This chapter qualitatively compares four neural network classifiers ELM, SRAN, BPNN and PNN for six ventricular arrhythmias. The PNN classifier has 99.8% accuracy, a stronger ECG classification accuracy than other classifiers.

REFERENCES

1. Bianca S. Honnekeri, Disha Lokhandwala, Gopi Krishna Panicker, Yash Lokhandwala, "Sudden cardiac death in India: A growing concern", *Journal of the Association of Physicians of India*, 62(12), 2014, pp. 36–40.
2. L. Goldberger, L. A. Amaral, L. Glass et al., "PhysioBank, PhysioToolkit, and Physionet: Components of a new research resource for complex physiologic signals", *Circulation*, 101(23), 2000, pp. E215–E220.
3. A. C. Lin, "Improved frequency-domain analysis of ventricular late potentials", *Computers in Cardiology*, 32(5), 2005, pp.479–482.
4. R. G. Hohlfeld, C. Rajagopalan, G. W. Neff, *Wavelet Signal Processing of Physiologic Waveforms*. Wavelet Technologies, Inc., 2004.
5. S. Z. Mahamoodabadi, A. Ahmedian, M. D. Abolhasani, "ECG feature extraction using daubechies wavelet", in *Proceedings of the Fifth IASTED International Conference Visualization, Imaging and Image Processing*, September 7–9, 2005, Benidorm, Spain.
6. R.Shantha, Selva Kumari, S. Bharathi, V. Sadasivam, "Design of optimal discrete wavelet for ECG Signal using orthogonal filter bank", in *International Conference on Computational Intelligence and Multimedia Applications*, IEEE, 2007.
7. David Cuesta-Frau, Daniel Novak, Vladimir Eck, Juan C. Pérez-Cortés, "Electrocardiogram baseline removal using wavelet approximations, computer-based medical systems", in *Proceedings of the 15th IEEE Symposium 2002*, CBMS, 2002.
8. N. Nikolaev, Z. Nikolov, A. Gotchev, K. Egiazarian, *Wavelet Domain Wiener Filtering for ECG Denoising Using Improved Signal Estimate*. IEEE, 2000.
9. G. Umamaheswar Reddy, M. Muralidhar, S. Varadarajan, "ECG De-Noising using improved thresholding based on Wavelet transform", *International Journal of Computer Science and Network Security*, 9(9), September 2009, pp. 221–225.
10. S. N. Shivappriya, R. Shanthaselvakumari, T. Gowrishankar. "ECG delineation using stationary wavelet transform", Advanced Communication and Computing, 2006,

ADCOM International Conference on Issue Date December 20–23, 2006. IEEE Xplore: August 13, 2007. DOI: 10.1109/ADCOM.2006.4289898.

11. S. N. Shivappriya, R. Harikumar, "A novel approach for different morphological characterization of ECG signal", *Lecture Notes in Electrical Engineering* 221, 2013. DOI: 10.1007/978-81-322-0997-3.

12. Dr. R. Harikumar, S. N. Shivappriya, "Feature extraction and classification of ECG signal proceedings", IETSD 2012: September 3–5, 2012, BIT Sathy, EC307–EC316.

13. R. Acharya, P. S. Bhat, S. Iyengar, "Classification of heart rate data using artificial neural network and fuzzy equivalence relation", *The Journal of the Pattern Recognition Society*, 36, 2002, pp. 61–68.

14. S. Suresh, S. Saraswathi, N. Sundararajan, "Performance enhancement of extreme learning machine for multi-category sparse data classification problems", *Engineering Applications of Artificial Intelligence Elsevier Journal*, 23(7), 2010, pp. 1149–1156.

15. Jinkwon Kim, Hang Sik Shin, "Robust algorithm for arrhythmia classification in ECG using Extreme Learning Machine", *BioMedical Engineering Online*, 8, 2009, p. 31. DOI: 10.1186/1475-925X-8-31.

16. B. S. Mahanand, S. Suresh, N. Sundararajan, M. Aswatha Kumar, "Disease detection using a self-adaptive resource allocation network classifier", *Proceedings of International Joint Conference on Neural Networks, San Jose, California, USA*, July 31–August 5, 2010.

17. P. M. Grant, "Artificial neural network and conventional approaches to filtering and pattern recognition", *Electronics and Communications Engineering Journal*, 1(5), 1989, pp. 225–232.

18. A. S. Al-Fahoum, I. Howitt, "Combined wave-let transformation a classifying life threatening cardiac arrhythmias", *Medical and Biological Engineering and Computing*, 37(5), 566–573, 1999.

19. H. Al-Nashash, "Cardiac arrhythmia classification using neural networks", *Technology and Healthcare*, 8(6), 2000, pp. 363–372.

20. D. F. Specht, "Probabilistic neural networks for classification mapping, or associative memory", *Proceedings IEEE International Conference on Neural Networks*, 1, 1988, pp. 525–532.

21. E. Pietka, "Feature extraction in computerized approach to the ECG analysis", *Pattern Recognition*, 24(2), 1991, pp. 139–146.

22. F. Jager, G. B. Moody, A. Taddei, "Performance measure for algorithms to detect transient ischemic ST segment changes", *Proceedings Computing in Cardiology*, 1991, pp. 369–372.

23. G. Subramanya Nayak, C. Puttamadappa, "Neural network based classification of ECG signals using LM algorithm", *World Academy of Science, Engineering and Technology*, 60, 2009, pp. 579–581.

15 Computer-Aided Diagnosis toward Improving Severity Assessment of Dysarthria

*Shanmugam Arun Kumar, Sivaraj Dhakshin,
Kumara Krishna Ananthan Viveka Vikram,
Krishnagoundenpudur Natarajan Anu Prabha,
and Shanmugam Sasikala*

CONTENTS

15.1 INTRODUCTION

Communication is an act of people expressing their ideas and thoughts through various methods. Humans exhibit through emotions, speaking, writing, and body language. The most frequent and habitual mode is speaking. Speaking is one of the most distinguished and gifted skills of human life [1]. But speech disorders affect the speaking capability of humans. One of the major speech disorders is found to be dysarthria. Dysarthria is a speech disorder that is caused mainly due to damage in the brain and nervous systems. Common causes of dysarthria include nervous disorders such as stroke, brain tumors, brain injury, and circumstances that originate with facial paralysis or tongue muscle flaw.

Speech is an expression of ideas in spoken words. It is characterized by articulation, pronunciation, pitch, rate, rhythm, and pauses. Any variation in these features may affect intelligibility. People with dysarthria will have problems controlling the pitch and loudness of his/her own speech. Various types of dysarthria include spastic, ataxic, flaccid, hypokinetic, hyperkinetic, and mixed. Initially, there is no detectable speech disorder or changes in speech performance. Then a detectable change in voice is identified, with a reduction in pitch or loudness. This leads to a limited range of oral movements. As a result of the severity and frequent communication breaks natural speech is no longer functional. Thus, quality of life is affected for people with speech disorders.

Speech pathologists check the patients to find out how severe the condition is. They test the ability to breathe and the oral movements of the mouth. Other tests include Magnetic Resonance Imaging (MRI) scans of the brain and neck and evaluation of swallowing capacity. This helps to identify whether they are related to the disorder but does not with the accuracy of what the type is being experienced. This requires automated techniques to classify the types with high accuracy to rectify the disorder.

In general, nervous system disorders completely affect the patient's lifestyle. Patients with neurological disorders face many issues in social life as they lack confidence. Dysarthric patients also face the same difficulty owing to their slurred speech and they may hesitate to communicate. Our work focuses on retaining the quality of life of dysarthric patients.

Speech processing in conjunction with Machine Learning helps in the accurate diagnosis and the automatic classification of the type of dysarthria.

The significant contributions in this work are:

- To classify the severity in dysarthria patients using Machine Learning. In addition, a performance comparison between different classifier techniques is presented to assess the severity.
- To provide an efficient identification of the severity of dysarthric patients.
- To assist the clinician in giving the exact treatment and speech therapy for dysarthric patients.

The proposed dysarthric speech severity assessment system is assessed in terms of accuracy, recall, precision, F1 score, Mathews correlation coefficient, and ROC area.

This chapter is organized as follows: Section 15.2 discusses the review of the literature. Section 15.3 describes the methodology employed. Section 15.4 discusses the results and inference. Section 15.5 focuses on the conclusion and future scope.

15.2   RELATED WORKS

Many works in the literature have focused on building systems for disordered voice assessment. However, only a few works have experimented on classifying the severity of dysarthria [2–4].

Minu et al. proposed a software tool to classify the severity of dysarthria. Pitch, formant frequency, jitter and shimmer-based features were used for classification [2].

In studies [3, 4] assistive systems for enhancing the intelligibility of dysarthric speech have been developed. The accuracy of severity assessment systems can be improved by enhancing dysarthric speech.

A mobile assistive system to classify normative speech with disordered speech was proposed in [5]. The performance of the system was tested on different Machine Learning algorithms. Correlation, information gain, and principal component-based feature selection have been used. A significant improvement in accuracy was achieved with the correlation method using a decision tree classifier.

A pathological voice classifier was proposed based on Mel-frequency cepstral coefficients (MFCC) and MFCC with delta features using SVM classifier on a 2018 FEMH voice data challenge database [6].

A novel disordered speech classifier was built using a feature selection algorithm with SVM and DT classifiers [7]. Dysphonia features with a feature vector of size 132 are used in the proposed study.

In [8], genetic algorithm-based feature selection was employed to classify Parkinson's disease within a healthy patient. Maximum vocal fundamental frequency, minimum vocal fundamental frequency, jitter, and shimmer features exhibited a high degree of correlation for discriminating between healthy and disordered populations.

Phonation, articulatory, and prosodic features were used for assessing the varying levels of severity of Parkinson's disease. A reasonable improvement in accuracy was obtained for binary classification [9].

A metric based on formant frequency and speech rate was proposed for assessing the severity of dysarthria. The proposed metric can be used by speech therapists for severity assessment [10].

A novel pathological voice classification system was proposed in [11] using wavelet features and the SVM classifier. The wavelet filter bank parameters are fine-tuned using a genetic algorithm to achieve high classification accuracy. The algorithm was tested on MIT-KAY database.

An automatic detection system to classify the severity of dysarthria using perceptually related speech features is experimented on in [12]. A comparative study between traditional and automatic severity assessment is also studied in detail.

Various features obtained from non-motor, imaging and anatomical biomarkers in conjunction with an ML algorithm are used for early diagnosis of Parkinson's disorder a type of degenerative dysarthria [13].

In [14], manual segmentation is done on the UCLASS stuttered database to analyze repetitions, prolongations, and interjections in speech utterances. Segmentation is followed by MFCC feature extraction and classification using K-NN and SVM classifiers to classify fluent and dysfluent speech.

15.3 MATERIALS AND METHODS

15.3.1 SPEECH CORPUS

The proposed work is tested on speech utterances from the Nemours database [15]. The database consists of 814 meaningless sentences each spoken by 11 male speakers with varying levels of severity. It also contains two connected speech paragraphs produced by them. Sentences used in the database are of the form, "The A is B-ing the C", where A and C are nouns and B is a verb. This database contains randomly

TABLE 15.1
Classification of Speakers in the Nemours Database

Speakers	Intelligibility score	Severity level
FB	92.9	Mild
MH	92.1	Mild
BB	89.7	Mild
LL	84.4	Mild
JF	78.5	Moderate
RL	73.3	Moderate
RK	68.6	Moderate
BK	58.2	Severe
BV	57.5	Severe
SC	51.5	Severe

generated sentences and sentences that include all pronunciations that could be made. The second 37 sentences are formed by swapping the nouns of the earlier set. The database consists of all possible sentences to test the communicative capability and severity level of dysarthric speakers. These phonemes are placed at the start or end of the word and at the middle position. On further process, phonemes are classified by voicing, primary articulator, and articulation manner.

The database consists of voice samples collected from persons with varying levels of severity. Table 15.1 shows the intelligibility scores and severity levels of 11 speakers in the Nemours database.

15.3.2 METHODOLOGY

The architecture of the proposed system is depicted in Figure 15.1. It summarizes the methodology involved in developing a dysarthric speech severity detection system. The first step is feature extraction. Acoustic and spectral features are extracted in the feature extraction phase. Feature extraction is followed by disease classification using a Machine Learning classifier. From the classification phase, the confusion matrix is computed and metrics such as accuracy, recall, precision, F1 score, Mathews correlation coefficient, and ROC area are used to evaluate the system performance. A comparative study on performance metrics on three different algorithms such as Support Vector Machine (SVM), Decision Tree (DT) and Multi-Layer Perceptron (MLP) used for classification is also presented. The proposed system could be used as an assistive rehabilitation device for dysarthric speakers and a diagnostic tool for clinicians to predict the severity of the disorder.

15.3.3 SPEECH ANALYSIS

Temporal, spectral, and prosodic features clearly discriminate the variations between different dysarthric speakers.

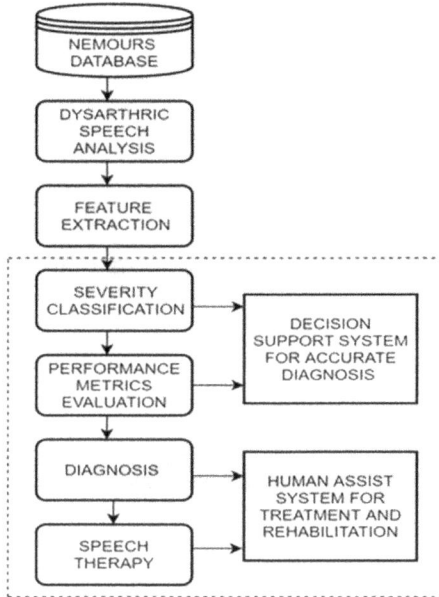

FIGURE 15.1 System architecture of dysarthric speech severity assessment system

On analyzing the spectral characteristics, there is an abrupt variation in the spectrum of dysarthric speech when compared to normative speech [3]. Time domain characteristics of dysarthric speech reveal longer duration and pauses between different words. The time domain plot and spectral plot of dysarthric and normative speech are shown in Figure 15.2 and Figure 15.3 respectively.

15.3.4 FEATURE EXTRACTION

Temporal features and spectral features are used for building a dysarthric speech severity system. Short-time energy, energy entropy and zero crossing rate are the temporal features [16]. Spectral centroid, spectral roll-off and spectral flux are the spectral features employed in the feature extraction phase [17].

- Short-time energy: Energy varies for voiced, unvoiced, and silent segments in the speech signal. Computation of short-time energy in speech ideally classifies variations in speaking styles of dysarthric patients.
- Zero crossing rate: The number of zero crossings in a voice in a speech waveform is denoted by the zero crossing rate. The zero crossing rate varies for different degrees of dysarthria.
- Energy entropy: Entropy represents the amount of information contained in a speech signal. Energy entropy corresponds to maximum entropy in energy-based regions of a speech signal.

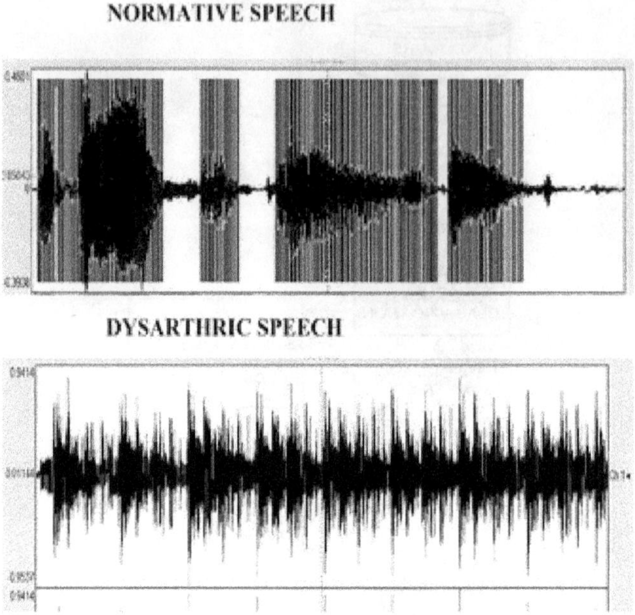

NORMATIVE SPEECH

DYSARTHRIC SPEECH

FIGURE 15.2   Time domain analysis of dysarthric and normative speech

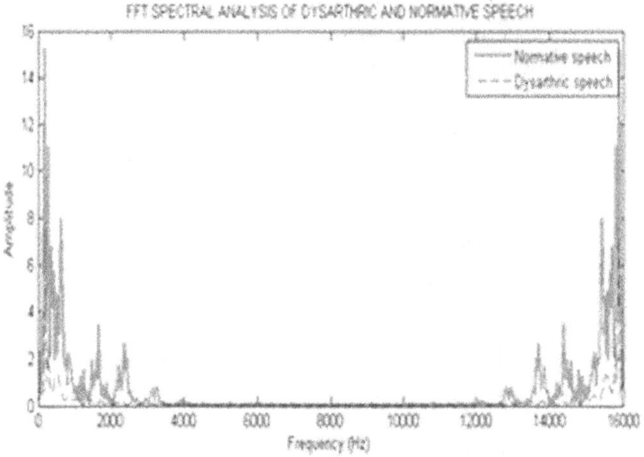

FIGURE 15.3   Spectral analysis of dysarthric and normative speech

- Spectral centroid: Spectral centroid corresponds to the center of the spectrum of the speech signal. A high value of spectral centroid specifies the concentration of energy at higher frequencies.
- Spectral roll-off: The spectral roll-off point measures the bandwidth at which a percentage of total energy exists.

- Spectral flux: Spectral flux is a metric which denotes the variability of the spectrum over time.

15.3.5 SEVERITY CLASSIFICATION

Feature extraction is followed by classification with a 10-fold cross validation. Any classification algorithm includes the training and testing phase. The training phase in an ML algorithm builds a model for classification. The testing phase evaluates the model on test data, predicts the label, and computes the confusion matrix. Metrics such as accuracy, recall, precision, F1 score, Mathews correlation coefficient, and ROC area are evaluated from the confusion matrix. SVM, DT and MLP are used to classify the severity of dysarthric speech, and a comparative study is performed.

Support vector machine: Support vector machine (SVM) is a supervised ML algorithm used widely in classification problems [18]. SVM is a maximum margin classifier that maximizes the distance between two classes of a hyperplane. The optimal hyperplane is built from the model generated using the training data. The model is used further to predict the output from the test data. SVM uses linear or non-linear kernels to classify data. If the data is linearly separable then linear SVM is used for classification. On contrary, highly non-linear data is classified based on kernel transformation. Kernel transformation linearizes non-linear data in higher dimensions. In this work, non-linear SVM (polynomial kernel) is used for classification.

Decision tree: A decision tree [19] is also a classifier model which models or built a tree structure to predict the output. It sub-divides the data into multiple data and checks every data and makes a decision and predicts the output.

Multi-layer perceptron: Multi-layer perceptron (MLP) [20] is a classical neural network comprising one or more neurons. Data is applied to the input layer and abstraction is achieved using multiple hidden layers. Predictions are made from the output layer of the neural network.

15.4 EXPERIMENTAL RESULTS AND DISCUSSIONS

Speech samples collected from speakers BB, BK, BV, FB, and JF of the Nemours database are used to develop a five-class severity detection. Feature extraction is followed by classification and the performance metrics such as accuracy, recall, precision, F1 score, Mathews correlation coefficient, and ROC area are determined from the confusion matrix. The mathematical equations relating to the metrics are explained in Figure 15.4.

From the performance metrics of Table 15.2, it is evident that the performance of the dysarthric severity detection system is significantly better for the multi-layer perceptron classifier and the decision tree classifier.

Accuracy is the percentage of correct classifications. An accuracy of 74.32% and 74.05% is achieved using temporal and spectral features. More than one performance metric is required to quantify the performance of a system. Henceforth other metrics such as precision, recall, F1 score, Mathews correlation coefficient (MCC) and ROC area are also calculated [21–23].

S. No.	Performance metric	Equation
1.	Accuracy	$ACC = \dfrac{TP+TN}{FP+FN+TP+TN}$
2.	Recall	$REC = \dfrac{TP}{TP+FN}$
3.	Precision	$PRE = \dfrac{TP}{TP+FP}$
4.	F1score	$F_1 = 2 \times \dfrac{PRE \times SEN}{PRE + SEN}$
5.	Matthews Correlation Coefficient	$MCC = \dfrac{TP.TN - FP.FN}{\sqrt{(TP+FP)(TP+FN)(TN+FP)(TN+FN)}}$

TP/FP - True/False positive
TN/FN- True/False Negative

FIGURE 15.4   Performance metrics computed from a confusion matrix

---

TABLE 15.2
Performance Metrics of Dysarthric Severity Classification System

Classifier	ACC %	PRE %	REC%	F1	MCC	ROC Area
SVM	70.81	71.8	70.8	0.706	0.639	0.872
DT	74.05	74.0	74.1	0.739	0.675	0.923
MLP	74.32	73.9	74.3	0.739	0.677	0.922

Precision indicates positive predictive value. A high value of precision corresponds to a low false positive. Precision is high and its value lies above 0.73.

Recall justifies that DT and MLP algorithms identify severity correctly 74% of the time.

The F1 score is related to the accuracy of the classifier. The F1 score is computed by striking a balance between false positives and false negatives. The F1 score is high in DT and MLP systems.

MCC is a measure of prediction performance. A value of 1 in MCC corresponds to exact prediction and 0 corresponds to inverse prediction. From the results, the MCC value is above 0.67 in both DT and MLP classifier algorithms.

The receiver operating characteristics (ROC) area gives the general performance of a classifier. The ROC area is high and lies above 0.92 for DT and MLP classifiers.

From the metrics, it is evident that all six-performance metrics are significantly higher for the MLP classifier. This accuracy is less because there is more fluctuation in the frequency domain of the dysarthric speech and to improve the accuracy highly discriminative features are to be considered.

15.5 CONCLUSION

In this chapter, the classification of dysarthria using features such as short-time energy, energy entropy, zero crossing rate, spectral centroid, spectral roll-off, and spectral flux have been evaluated. This classification technique helps doctors to identify the exact type of dysarthria. This work will assist the doctor to give the exact treatment and therapy for dysarthric patients.

To further improve the accuracy in the multiclass, multiple features with a feature selection may be employed. The future scope of this work is toward improving the system performance using high-level feature transformation techniques, feature selection, and evolutionary algorithms. A deep learning framework may also be employed for enhanced diagnosis and assessment.

REFERENCES

1. O'Shaughnessy, Douglas. *Speech Communication: Human and Machine.* Universities Press, 1987.
2. Thoppil, Minu George, C. Santhosh Kumar, and John Amose, Anand Kumar. "Speech signal analysis and pattern recognition in diagnosis of dysarthria," *Annals of Indian Academy of Neurology* 20(4) (2017): 352.
3. Arun Kumar, S., and C. Santhosh Kumar. "Improving the intelligibility of dysarthric speech towards enhancing the effectiveness of speech therapy," in *2016 International Conference on Advances in Computing, Communications, and Informatics (ICACCI),* pp. 1000–1005, IEEE, 2016.
4. Arun Kumar, S., K. Kavitha, and A. Kumaresan. "Intelligible transformation techniques towards enhancing the intelligibility of dysarthric speech: A review," *International Journal of Pure and Applied Mathematics* 118(7) (2018): 573–80.
5. Verde, Laura, Giuseppe De Pietro, and Giovanna Sannino. "Voice disorder identification by using machine learning techniques," *IEEE Access* 6 (2018): 16246–16255.
6. Pishgar, Maryam, et al. "Pathological voice classification using mel-cepstrum vectors and support vector machine,"in *2018 International Conference on Big Data,* pp. 5267–5271, IEEE, 2018.
7. Tsanas, Athanasios, et al. "Novel speech signal processing algorithms for high-accuracy classification of Parkinson's disease," *IEEE Transactions on Bio-Medical Engineering* 59(5) (2012): 1264–1271.
8. Shahbakhi, Mohammad, Danial Taheri Far, and Ehsan Tahami. "Speech analysis for diagnosis of Parkinsons disease using genetic algorithm and support vector machine," *Journal of Biomedical Science and Engineering* 7(4) (2014): 147–156.
9. Khan, Taha, Jerker Westin, and Mark Dougherty. "Classification of speech intelligibility in Parkinson's disease," *Biocybernetics and Biomedical Engineering* 34(1) (2014): 35–45.
10. Simm, W. A., P. E. Roberts, and Malcolm J. Joyce. "Signal processing for use in the assessment of dysarthric speech," in *3rd IEE International Seminar on Medical Applications of Signal Processing 2005* (Ref. No. 2005-1119). IET, 2005.

11. Saeedi, Nafise Erfanian, Farshad Almasganj, and Farhad Torabinejad. "Support vector wavelet adaptation for pathological voice assessment," *Computers in Biology and Medicine* 41(9) (2011): 822–828.
12. Guerra, E. Castillo, and Denis F. Lovey. "A modern approach to dysarthria classification," in *Proceedings of the 25th Annual International Conference of the IEEE Engineering in Medicine and Biology Society* (Vol. 3), pp. 2257–2260, IEEE, 2003.
13. Prashanth, R., et al. "High-accuracy detection of early Parkinson's disease through multimodal features and machine learning," *International Journal of Medical Informatics* 90 (2016): 13–21.
14. Mahesha, P., and D. S. Vinod. "An approach for classification of dysfluent and fluent speech using K-NN and SVM," *International Journal of Computer Science, Engineering and Applications* 2(6) (2012): 23–32.
15. Menendez-Pidal, Xavier, James B. Polikoff, Shirley M. Peters, Jennie E. Leonzio, and H. T. Bunnel. "The nemours database of dysarthric speech," *ICSLP 96: Proceedings of the Fourth International Conference on Spoken Language*, Vol. 3, pp. 1962–1965, 1996.
16. Jalil, Madiha, Faran Awais Butt, and Ahmed Malik. "Short-time energy, magnitude, zero crossing rate and autocorrelation measurement for discriminating voiced and unvoiced segments of speech signals," in *International Conference on Technological Advances in Electrical, Electronics and Computer Engineering (TAEECE)*, pp. 208–212, IEEE, 2013.
17. Alas, Francesc, Joan Claudi Socor, and Xavier Sevillano. "A review of physical and perceptual feature extraction techniques for speech, music and environmental sounds," *Applied Sciences* (2016): 143.
18. Boser, Bernhard E., Isabelle M. Guyon, and Vladimir N. Vapnik. "A training algorithm for optimal margin classifiers," in *Proceedings of the Fifth Annual Workshop on Computational Learning Theory*, pp. 144–152, ACM, 1992.
19. Quinlan, J. Ross. "Induction of decision trees," *Machine Learning* 1 (1986): 81–106.
20. Murtagh, Fionn. "Multilayer perceptrons for classification and regression," *Neurocomputing* 2 (5–6)(1991): 183–197.
21. Sasikala, S., et al. "Fusion of MLO and CC view binary patterns to improve the performance of breast cancer diagnosis," *Current Medical Imaging Reviews* 14(4) (2018): 651–658.
22. Seliya, Naeem, Taghi M. Khoshgoftaar, and Jason Van Hulse. "A study on the relationships of classifier performance metrics," in *IEEE International Conference on Tools with Artificial Intelligence*, pp. 59–66, IEEE, 2009.
23. Arun Kumar, S., and S. Sasikala. "Towards enhancing the performance of a stress detection system," *International Journal of Innovative Technology and Exploring Engineering (IJITEE)* 8(2S) (2018): 379–383.

16 SVM-Based Demodulator for QPSK Signal Suitable for Vehicular Communication

*Kanakaraj Kavitha, Kumaraswamy Jasmine,
Shanmugam Arun Kumar, and
Ramasamy Dhivya Praba*

CONTENTS

16.1 INTRODUCTION

Vehicular communication systems gain attraction among industry and academia due to Intelligent Transport Systems (ITS) concepts used in the smart city models to implement smart transportation. The different types of vehicular communication [1] scenarios are Vehicle-to-Vehicle (V2V), Vehicle-to-Pedestrian (V2P) and Vehicle-to-Infrastructure (V2I). Whatever the situation, non-zero relative velocity between the transmitter and the receiver is inherent in the situation which results in significant Doppler spread, which would be a major concern. Further, it also includes a multipath fading effect. The modulation techniques used in standard wireless technologies, which may be the choice for the ITS implementations need to be investigated for vehicular channel conditions. WAVE (Wireless Access in Vehicular Environment) is a recently approved wireless standard suitable for the licensed ITS band of 5.9 GHz [2]. To meet the required reliability and safety it is needed to design robust physical layer technologies suitable for the vehicular channel [3].

MIMO (Multiple Input Multiple Output) and OFDM (Orthogonal Frequency Division Multiplexing) gain considerable attention in 5G and next-generation standards due to their advantages. Different diversity and capacity enhancement MIMO transmission techniques, such as space-time coding, spatial multiplexing, index modulation and hybridization of the techniques have been proposed and analyzed in the literature [4] for the last two decades. However, index modulation gains attention when vehicular communication is considered. Various index modulation techniques along with OFDM and space-time coding (STC) have been proposed and analyzed over various fading channel conditions in the literature [5–9]. Hasan et al. [10], propose an adaptive OFDM modulation scheme for an optical channel. However, all the above ideas give an improvement with the increase in system and computational complexity [11].

The Machine Learning approach is explored for signal classification in papers [12, 13]. The simulated results elaborated in these papers show that the SVM-based signal classification could achieve improvement over fading channels. Ahmad et al. [14] in their paper have proposed a deep learning convolutional neural network (DCNN) approach for BFSK demodulation over fading channel and compared with SVM, multilayer perceptron (MLP), quadratic discriminant analysis (QDS), linear discriminant analysis (LDA) and conventional demodulator. They have also analyzed the computational complexity issues and solutions for optimum demodulation. The QPSK is a bandwidth-efficient signaling scheme and performs very poorly over a hasty wireless channel with conventional coherent detection. Machine signal classification may be suitable for such situations. SVM is a powerful and simple classification algorithm which has been used widely for many applications. It is a hyperplane-based discriminative classifier which works well when the data has maximum randomness [15, 16].

In this chapter, an SVM-based signal classification for the QPSK signal has been proposed and analyzed over slow and fast fading channels using simulation tools. The content of this chapter has been organized as below: Section 16.2 describes the system model and Section 16.3 gives the simulation results and analysis. Section 16.4 gives the conclusion.

16.2   SYSTEM MODEL

Figure 16.2 shows the transmitter and receiver configurations for the proposed SVM-based signal classifier. The binary data sequence b is given to the QPSK modulator and the modulated signal $x(t)$ is transmitted over the channel.

$$r(t) = g(t) \times (t) + n(t)$$

Where, $g(t)$ is the path gain and $n(t)$ is the additive white Gaussian noise (AWGN) process. When the AWGN channel is considered the channel gain factor $g(t)$ will be 1. The received signal $r(t)$ is given to the correlation receiver as described in Figure 16.1. It consists of a multiplier followed by an integrator. The output of the coherent receiver r is given to the maximum likelihood–based hard decision device in the conventional demodulator. In the SVM-based signal classifier, the correlation receiver output is given to the SVM classifier. The SVM classifier is trained with the known sequence of length N transmitted as the first part of the frame. After training, it classifies the date received in the rest of the frame (Figure 16.2).

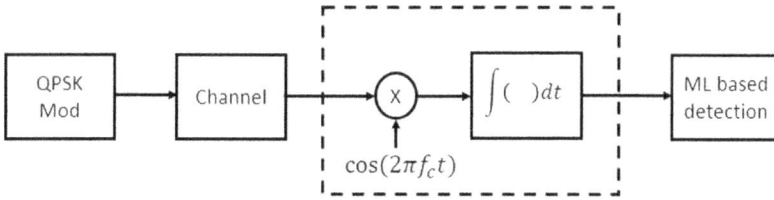

FIGURE 16.1   Transmitter and receiver configuration for a conventional QPSK demodulator

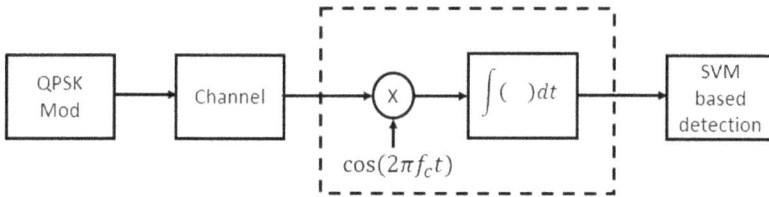

FIGURE 16.2   Transmitter and receiver configuration for SVM-based demodulator

16.3 SIMULATION RESULTS

In this section the BER performance of the SVM-based demodulator for a QPSK signal is compared with the conventional demodulator. The block length of 1000 bits with 100 training bits is considered for SVM-based decoding. The BER performance for the signal-to-noise ratio range considered is obtained by averaging the error rate for 10 frames.

Figure 16.3 shows the BER performance of the QPSK signal over AWGN with a conventional demodulator and the SVM-based demodulator. The graph shows that the SVM needs about 2dB SNR to achieve the BER of 10^{-3} compared to the AWGN channel.

The BER performance of the SVM-based and conventional demodulators have been compared in the Figure 16.4 over slow and flat Rayleigh fading channels. The results reveal that with sufficient training the SVM classifier could able to demodulate the signal over slow and flat Rayleigh fading channel compared to the AWGN channel. This may be because of the diversity introduced due to the Rayleigh fading in the training sequence, which would help the SVM learning be effective compared to the AWGN channel.

In Figure 16.5, the SVM-based demodulator is compared over the fast and slow Rayleigh fading channel and the AWGN channel. The SVM-based decoder shows equivalent performance in all three-channel conditions.

16.4 CONCLUSION

In this chapter, an SVM-based demodulator approach is proposed for the QPSK signal suitable for vehicular communication. The results show that the SVM-based

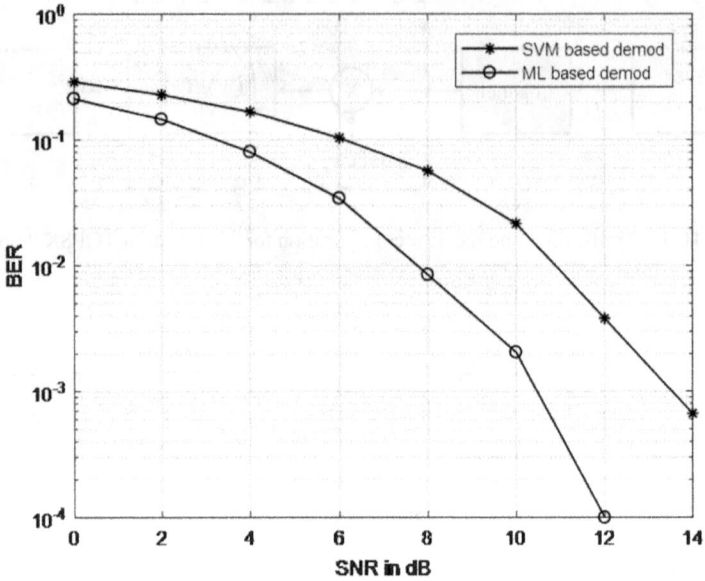

FIGURE 16.3 BER performance of the SVM-based demodulator and conventional ML-based demodulator over AWGN channel for the QPSK signal

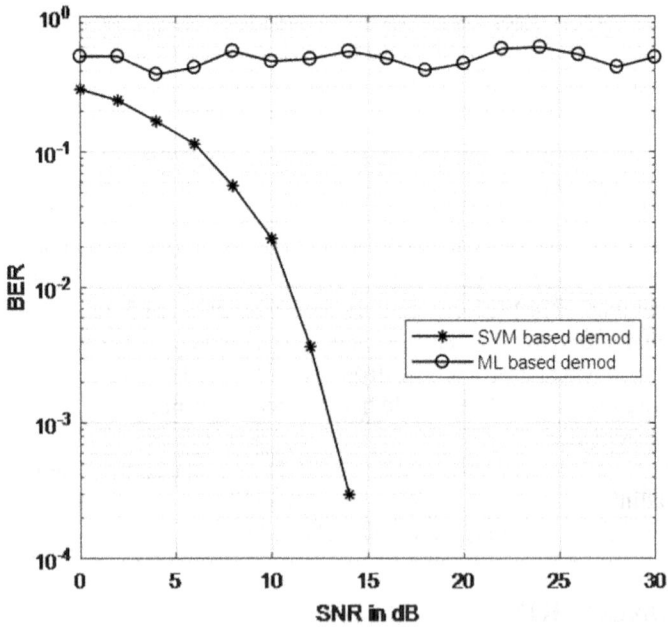

FIGURE 16.4 BER performance of the SVM-based demodulator and conventional ML-based demodulator over flat and slow Rayleigh fading channel for the QPSK signal

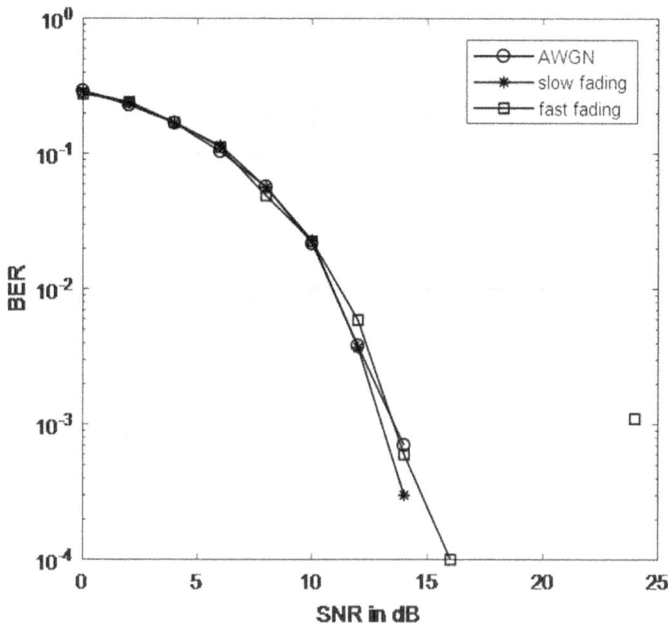

FIGURE 16.5 BER performance of the SVM-based demodulator and conventional ML-based demodulator over fast and slow Rayleigh fading channel for the QPSK signal

demodulator is outperforming the conventional demodulator over fading channel conditions compared to the AWGN channel. However, the improvement is achieved with the cost of the training sequence and complexity. The simulation results show that over Rayleigh fading the SVM-based decoder outperforms the conventional demodulator. Further, the SVM shows equivalent performance irrespective of channel condition. Hence, it may be suitable for unpredictable vehicular channel.

REFERENCES

1. Viriyasitavat, Wantanee, et al. "Vehicular communications: Survey and challenges of channel and propagation models." *IEEE Vehicular Technology Magazine* 10(2) (2015): 55–66.
2. Ma, X., et al. "The physical layer of the IEEE 802.11 p WAVE communication standard: The specifications and challenges." *Electric Vehicles Model Simulations* 2(3) (2014): 1–5.
3. Jasmine, K. "Performance analysis of diversity techniques for wireless communication." *International Journal of Advanced Research in Electrical, Electronics and Instrumentation Engineering (IJAREEIE)* 9(5) (2020): 1088–1093.
4. Kavitha, K., and H. Mangalam. "'Multilevel coding for MIMO' book chapter, computational intelligence and sustainable systems: Intelligence and sustainable computing." Springer, 2019.
5. Kavitha, K., and H. Mangalam. "Multilevel spatial modulation." *Journal of the Chinese Institute of Engineers* 39(6) (2016): 713–721.

6. Kavitha, K., and H. Mangalam. "Multilevel spatial multiplexing–space time trellis coded modulation system for fast fading MIMO Channel." *International Journal of Engineering and Technology* 6(1) (2014): 217–222.
7. Kavitha, K., and H. Mangalam. "Low complex decoding algorithm for multilevel space time trellis codes over MIMO channel." *International Journal of Information and Communication Technology* 8(1) (2016): 69–78.
8. Kavitha, K., A. Kumaresan, and S. Arun Kumar. "Performance analysis of multilevel spatial modulation OFDM technique (MLSM- MIMO)." *International Journal of Pure and Applied Mathematics* 116(11) (2017): 101–109.
9. Kavitha, K., and N. M. Harshini. "Noma-Stbcfor uplink Mimo Channel. " *International Journal of Engineering and Advanced Technology* 8(6S3) (2019): 1540–1542.
10. Farahneh, Hasan, Fatima Hussain, and Xavier Fernando. "Performance analysis of adaptive OFDM modulation scheme in VLC vehicular communication network in realistic noise environment." *EURASIP Journal on Wireless Communications and Networking* 2018(1) (2018): 243.
11. Yang, Ping, et al. "Multi-domain index modulation for vehicular and railway communications: A survey." IEEE, 2018.
12. Dubois, J. P., and Omar M. Abdul-Latif. "Least square-SVM detector for wireless BPSK in multi-environmental noise." *World Academy of Science, Engineering and Technology, International Journal of Electrical, Computer, Energetic, Electronic and Communication Engineering* 2(8) (2008): 1692–1697.
13. Dubois, Jean-Pierre, and Omar M. Abdul-Latif. *Improved M-Ary Signal Detection Using Support Vector Machine Classifiers.* IEC, 2005.
14. Mohammad, Ahmad Saeed, et al. "Demodulation of faded wireless signals using deep convolutional neural networks." *2018 IEEE 8th Annual Computing and Communication Workshop and Conference (CCWC).* IEEE, 2018.
15. Cortes, Corinna, and Vladimir Vapnik. "Support-vector networks." *Machine Learning* 20(3) (1995): 273–297.
16. Alpaydin, Etherm. *Introduction to Machine Learning*, 2nd ed., PHI Learning Private Limited, 2015.

17 Medicinal Image Watermarking System for Recovering Embedded Information from Therapeutic Restorative Images

S. Uma Maheswari and C. Vasanthanayaki

CONTENTS

17.1 INTRODUCTION

The exchange of helpful images between recuperation offices arranged at remote spots has transformed into a routine in present times. This exchange of helpful images brings about two confinements for the restorative images: the information has not been changed by unapproved people and there should be confirmation that the information is with the correct patient [1]. On the other hand, the transmission of patient data and its helpful images shared freely through business frameworks such as the Internet achieves unreasonable transmission time and cost. Watermarking is one of the procedures used to deal with more than two concerns.

DOI: 10.1201/9781003307716-17

Watermarking strategies have been described into two classes to be explicit spatial space and repeat zone. This game plan relies upon the medium used for concealing the data in an image. In spatial space watermarking methodologies [7, 10, 11, 13], data is embedded directly into the image while data is inserted into a changed host image in repeat space watermarking systems [8, 9, 14]. Another request for the watermarking procedure is to make the procedure both reversible and irreversible. In the reversible watermarking methodology [9, 14, 17, 18], the host image can be recovered accurately at the beneficiary's side from the watermarked image. The exact recovery of the host image isn't possible if there should be an event of irreversible watermarking techniques [8, 15]. Reversible watermarking is increasingly suitable for restorative images [2].

Four sorts of watermarking methods are used to guarantee propelled images: robust watermarking [3], fragile watermarking [4], semi-fragile watermarking [5] and hybrid watermarking [6]. Ground-breaking watermarking methodologies are used for copyright affirmation of electronic images as it is difficult to remove watermarks from cutting-edge images. Solid watermarks bolster intentional or unplanned strikes such as scaling, weight, altering, and so forth. Sensitive watermarking frameworks are ideal to check the confirmation of cutting-edge images. Any change or adjustment removes the fragile watermark from the watermarked image. Thusly, the non-attendance of the watermark demonstrates that the image has been adjusted. Semi-fragile watermarks endure incidental attacks. Cross-breed watermarks are the amalgamation of sensitive and solid watermarks. Here, robust watermarks are used for security control and sensitive watermarks are used for the uprightness control of the propelled image.

Most restorative images contain two areas called ROI and RONI. From a discovery point-of-view the ROI part is increasingly basic. Care should be taken while discovering data to the ROI part with the objective that visual quality won't be debased. Meanwhile, any modification to the ROI must be recognized and the primary ROI must be recovered remembering the ultimate objective is to avoid misdiagnosis and the retransmission of a remedial image. The recovery data of ROI is generally embedded into RONI [10, 11, 13–16, 19]. When any modification is recognized inside the ROI of the returned watermarked therapeutic image then the adjusted zone of ROI is replaced with the recovery data introduced inside RONI.

This paper proposes a novel piece-based sensitive remedial image watermarking system to achieve as per the goal.

1. Perceiving the proximity of the adjustment inside the ROI
2. Recovering the main ROI when it is changed
3. Perceiving modifications inside the ROI and recovering extraordinary ROI using irrelevant size affirmation and recovery data
4. Avoiding the route toward checking the ROI of the watermarked therapeutic image for the proximity of modifications when the ROI isn't adjusted
5. Embedding the EPR of patients into the remedial image

17.2   LITERATURE REVIEW

A number of watermarking techniques have been created to recognize changes in the ROI or in the entire watermarked restorative image and recover the primary ROI or the entire remedial image. Zain et al. [7] proposed a piece-based arrangement, the helpful image is divided into 8×8 squares and later mapping is set up between the squares for embedding the recovery information of each square into its related mapped piece. A short time later, each piece is furthermore parceled into four sub-squares of 4×4 size each and a while later a 9-bit watermark is made for each sub-piece. The made 9-bit watermark of each sub-piece is introduced into LSBs of the starting 9 pixels of the sub-block in the related mapped square. On the beneficiary's end, the watermarked therapeutic image is parceled into squares of 8×8 size and after that, the mapping between the pieces is found out through the embedding strategy. Subsequently, each piece is also isolated into four sub-squares of 4×4 size and after that, a 2-level disclosure plot is associated with recognizing modified squares. This 2-level distinguishing proof perceives adjusted pieces. Where 1-level disclosure is associated with sub-squares of pieces and the 2-level area is associated with squares. Right when a modified piece is perceived, the related mapped square is recognized and after that, the recovery data embedded in the mapped piece is isolated. This recovery data is used to supersede the pixels in a modified piece. Noteworthy drawbacks of this methodology are (1) in the event that both piece A and its mapped square B are changed, at that point it isn't possible to recover one of a kind image, (2) This system isn't using any confirmation data for the entire helpful image to check explicitly whether the image is modified. In this manner, all squares in the image must be checked one after the other to recognize the proximity of modifications. This checking system prompts wastage of time when the image isn't adjusted, (3) There is no course of action for introducing the EPR of a patient into the helpful image.

Wu et al. [8] made two piece-based procedures. In the subsequent system, a JPEG bit-string of the picked ROI is made and a short time later is isolated into settled length partitions. A short time later, the restorative image is confined into pieces and after that hash bits are processed for each square except the piece with ROI. These hash bits are used as affirmation data of the pieces. In each square of the image, hash bits of the piece and one part of the JPEG bit-string of ROI are both embedded using a solid-included substance watermarking system. By then all squares are combined to get the watermarked helpful image. At the beneficiary's end, the watermarked remedial image is divided into squares as done in the introduction strategy. From each square, hash bits of the piece and a section of the JPEG bit-string are both removed. For each square, hash bits are figured and after that appear differently in relation to the removed hash bits which check whether the piece is adjusted or not. If the square with ROI is perceived as adjusted, then the JPEG bit-string sections removed from all pieces are used to recover the ROI. Drawbacks of this method are (1) it isn't possible to get exceptional ROI as the JPEG bit-string of ROI is used to recover ROI when it is disturbed, (2) this technique requires a progressive number of calculations to create recovery data of ROI and introducing it into all bits of the helpful image, (3) the proportion of confirmation data is broad; for each square 150 bits are used, (4) there is no game plan for embeddings EPR of patients into the restorative image.

Chiang et al. [9] proposed a two-piece build methodology arranged in light of a symmetric key cryptosystem and changed the complexity improvement (DE) technique. The principle system can recover the whole therapeutic image, whereas the subsequent procedure can recover only the ROI of the restorative image. In the principle system, the remedial image is isolated into 4×4 size squares and subsequently a typical part of each piece is processed. Subsequently, the midpoints of all squares are connected and thereafter encoded using two symmetric keys k1 and k2 in order to extend the degree of security. By then, Haar wavelet change is associated on all squares to perceive smooth pieces. The encoded midpoints of the extensive number of squares are embedded in the recognized smooth pieces. At the authority's end, the embedded data is expelled from watermarked images and after that decoded using the keys k1 and k2 to get the midpoints, everything being equal. Thereafter, midpoints are learned for all squares and a short time later diverged from removed midpoints with recognized modified pieces. Exactly when a modified square is perceived then the pixels in the changed piece are displaced with the isolated typical of that square. The subsequent system is the same as the essential strategy except that the bits of pixels in squares of the ROI are embedded instead of the midpoints of all pieces in the entire image. Issues with these plans are 1) in the second system the proportion of approval and recovery data is far-reaching; 128 bits for each piece in the ROI, 2) the two methodologies require a greater open door for embedding data into the remedial image, as all squares of the therapeutic image must be changed into repeat territory and after that smooth squares must be recognized for introducing data, 3) the two strategies are not using any confirmation data for the entire ROI or the entire image to check directly whether the ROI or the entire image is adjusted. Along these lines, all pieces in the ROI or in the entire image must be checked in a relentless movement to perceive the proximity of adjustments. This checking system promotes wasting of time when the image isn't changed, 4) there is no game plan for introducing the EPR of patients into the remedial image.

Liew et al. [10, 11] made two reversible piece-based methodologies. In the primary method, the therapeutic image is partitioned into two territories: ROI and RONI. A short time later, ROI and RONI are apportioned into non-covering bits of sizes 8×8 and 6×6 exclusively. By then, a mapping is surrounded between squares of ROI to embed recovery information of each piece into its mapped piece. Each piece in ROI is mapped to a square in RONI. This mapping is used to introduce LSBs of pixels in an ROI block into its mapped RONI piece.

By then, the method proposed by Zain [7] is associated with just the ROI part to distinguish adjustments inside the ROI and recover the one-of-a-kind ROI. The first LSBs that were embedded in RONI are isolated and after that restored to their circumstances to get the principal remedial image. This procedure has an undefined weakness from the strategies proposed by Liew et al. [10, 11]. Qershi et al. [14] developed a reversible ROI-based watermarking plan. At the sender's end, the remedial image is assigned to both ROI and RONI. A while later, the patient's data and hash estimation of ROI are both introduced into ROI using the method suggested by Gou et al [1]. Stuffed sort of ROI, ordinary estimations of pieces inside ROI, embedding map for ROI, embedding map for RONI and LSBs of pixels in an emanate zone of RONI are introduced into RONI using the strategy of Atroni [22] Finally, information on ROI is embedded into LSBs of pixels in emanate locale. At the beneficiary's end, ROI information is expelled from the discharge locale and is used to recognize ROI and RONI areas. From the perceived RONI area stuffed sort of ROI, typical estimations of squares inside ROI, the embedding aide of ROI, introducing aide of RONI, and LSB of pixels in emanate districts are expelled. Using the expelled region guide of ROI, the patient's data and hash estimation of ROI are removed from ROI. By then, the hash estimation of ROI is determined and differentiated and isolated hash regard. If there is a bungle between the two hash regards, then the ROI is isolated into 16×16 pieces. For each square, the ordinary regard is learned and differentiated and the looking at typical motivation in the isolated environment. If they are not identical then the square is separate as adjusted and superseded by the relating bit of the compacted sort of ROI. Two weights of this procedure are (1) isolating the embedded data from RONI without realizing the embedding aid of RONI, (2) use of a compacted sort of ROI as recovery data for the ROI.

Al-Qershi and Khoo [15] proposed a game plan in light of two-dimensional refinement improvement (2D-DE). At the sender's end, the accommodating image is restricted into three regions: ROI pixels, RONI pixels, and fringe pixels. A brief span later, the association of the patient's information, hash estimation of ROI, bits of pixels inside ROI and LSBs of outskirts pixels are squeezed utilizing Huffman coding and in this way brought into RONI utilizing 2D-DE system. This presentation makes a strict design that will be associated with data of ROI and a brief timeframe later installed into LSBs of outskirts pixels. At the gatherer's end, from edge pixels in the watermarked medicinal image, the two data of ROI and zone plan were expelled. Utilizing this ROI data, ROI, and RONI are seen. The evacuated zone diagram is used to separate the patient's information, hash estimation of ROI, bits of pixels inside ROI and LSBs of edge pixels from RONI. The system for perceiving altered squares is the same as the one utilized as a bit of [14]. Each changed piece is supplanted by the differentiating square of pixels in the expelled ROI. The LSBs of edge pixels are supplanted utilizing the expelled LSBs from RONI. A basic issue of this plan is it is material to just the helpful images whose ROI measure is lower (up to 12% of the size of the whole image).

Al-Qershi and Khoo [16] developed a cross-breed ROI-based strategy. At the sender's end, the therapeutic image is detached into three areas: ROI, RONI, and edge pixels. Subsequently, the patient's data and hash estimation of ROI are embedded inside ROI using the changed DE system. The ROI zone plot with a stuffed kind of ROI and ordinary forces of pieces inside ROI are then introduced into RONI using the DWT framework. By then, the size of the watermark that is inserted into RONI and ROI information is embedded inside edge pixels using the equivalent

DWT framework. At the authority's end, ROI information is isolated from periphery pixels and is used to perceive both ROI and RONI regions. Pressed sort of ROI, ordinary forces of pieces in ROI and region guide of ROI are expelled from the recognized RONI area. Using the isolated zone guide of ROI, the patient's data and hash estimation of ROI are removed from ROI. The framework for perceiving modified squares and recovering ROI is the same as in [14]. Two weights of this procedure are (1) usage of stuffed kind of ROI as recovery information for the ROI, (2) relevant to simple images whose size is not under 512×512.

Deng et al. [17] developed a region-based adjusting recognizable proof and recovering method in perspective on reversible watermarking and quad-tree rot. In this procedure, an exceptional image is segregated into upsets with high homogeneity using quad-tree crumbling and after that, a recovery feature is figured for each square using a straight contribution of pixels. The recovery features of all squares are embedded as the first watermark using invertible number change. Quad-tree information as a second layer watermark is embedded using LSB substitution. In the approval arrangement, the embedded watermark is evacuated and the principal image is recovered. A similar straight acquaintance methodology is utilized to get each square's part. Adjusting ID and containment can be practiced by differentiating the isolated component and the recomputed one. The evacuated part can be used to recover those modified zones with a high likeness to their remarkable state. One drawback of this arrangement is the right exceptional image can't be recovered when it is changed.

17.3 PROPOSED METHOD

To accomplish the previously mentioned goals, this chapter proposes a medicinal image watermarking procedure.

17.3.1 DIVISION OF MEDICAL IMAGES

In a therapeutic image, the ROI is the most essential part to make a finding. A therapeutic image may contain a couple of disjoint ROI districts and may be in different shapes. The ROI parts are separated by a specialist or by a clinician shrewdly. Each roust zone is addressed by an encasing polygon. The encasing polygon is depicted by the number of vertices and their bearings. In the proposed procedure, the therapeutic image is separated into three areas of pixels: ROI pixels, RONI pixels. The restorative images containing a single ROI are used. The proposed strategy can in like manner be used with helpful images containing various ROI regions [12].

17.3.2 HASH ESTIMATION OF ROI

In the wake of picking the ROI, the hash estimation of the ROI is found using the cryptographic hash work MD5. This limit creates an extraordinary code for any data and is a confined limit. Choosing the commitment from the code made by MD5 isn't possible. The figured hash estimation of ROI is used to confirm ROI.

17.3.3   RUN LENGTH ENCODING

Ensuing to figuring hash estimation of ROI, the ROI is disconnected into non-covering bits of size 4×4, and after that typical regard is learned for each square. A watermark is delivered by connecting the hash estimation of ROI, LSBs of pixels inside ROI and EPR of patients. This made watermark is compacted using RLE. RLE is a clear lossless weight technique and is used to decrease the degree of the watermark. Exceptional data can be reproduced decisively from the pressed data. In this procedure, on the off chance that a bit is repeating a number of times in progression, at that point that bit course of action is replaced by a check regard and the bit. For example, the twofold data 000001111110000000 will be deciphered as five 0s, six 1s and seven 0s, and it is coded as (101, 0), (110, 1) and (111, 0). The primary matched data containing 18 bits is compacted to 12 bits. The stuffed watermark is mixed using a secret key k1 to give security. The resultant watermark is introduced into LSBs of ROI pixels.

17.3.4   MAPPING BETWEEN PIECES OF ROI AND RONI

In the wake of introducing a watermark in LSBs of ROI pixels, RONI is apportioned into non-covering bits of size 3×3 pixels. Expecting that the amount of squares inside ROI isn't as much as the number of pieces inside RONI, for each piece in ROI, the looked at mapped hinder in RONI is perceived using Equation (17.1).

$$B_{\mathrm{RONI}} = \left[k \times B_{\mathrm{ROI}} \mathrm{nod}\ N_b \right] + 1 \tag{17.1}$$

Where N_b is the number of squares in ROI, B_{RONI} is the piece number in RONI, k is a release key and is a prime number in the region of 1 and N_b, B_{ROI} is a square number in ROI. In the wake of mapping each rows piece to a RONI block, the typical estimation of each rows square is embedded inside the related mapped RONI piece.

Directly, the low-down embedding computation is explained as takes after.

17.3.5   EMBEDDING ALGORITHM

1. Divide the main helpful image into three areas: ROI pixels, RONI pixels, and edge pixels.
2. Discover hash regard (h1) of ROI using MD5.
3. Segment the pixels inside ROI into non-covering squares of size 4×4 each.
4. For each row square, discover typical regard and use it as approval and recovery data of that piece.
5. Assemble the least significant bits (LBSs) of all pixels inside ROI and mean this gathering as B.
6. Address the characters in the EPR of patients using ASCII code and after that get the twofold similarity.
7. Produce watermark w by connecting h1, B, and E.

8. Pack watermark w using RLE weight technique to make w_{comp}.
9. Scramble the watermark w_{comp} using a secret key k1.
10. Embed the bits of the mixed watermark into LSBs of pixels inside ROI.
11. Detachment RONI into non-covering bits of size 3×3 each.
12. Tolerating that the amount of squares in ROI isn't as much as the number of pieces in RONI, the plot impedes in ROI to a square in RONI using Equation (17.1).
13. For each row square, discover the ordinary power regard and after that supplement into LSBs of introductory 8 pixels in the mapped RONI piece.
14. Scramble the bits demonstrating the information of ROI using secret key k1.
15. Embed the mixed bits into the LSBs of edge pixels.

17.3.6 EXTRACTION ALGORITHM

1. Concentrate the mixed bits from the LSBs of periphery pixels in the watermarked remedial image.
2. Decode the removed bits to get information on ROI.
3. Recognize ROI pixels and RONI pixels in the watermarked remedial image.
4. Concentrate the encoded watermark from the LSBs of pixels.
5. Translate the removed watermark to get w_{comp}.
6. Decompress the w_{comp} to procure the hash regard (h1) of ROI, LSBs (B) of pixels inside ROI and EPR (E) of the patient.
7. Supersede the LSBs of pixels inside ROI with the bits in B.
8. Process hash regard (h2) of the ROI using MD5.
9. Difference h1 and h2. If h1=h2 then the ROI is veritable and the extraction system closes.
10. If h1≠h2 then the ROI isn't genuine and is adjusted. Keep on following the stage to distinguish modified squares inside ROI and recover one-of-a-kind ROI.
11. Detachment ROI and RONI into bits of size 4×4 and 3×3 exclusively. For each line piece perceive the mapped RONI square using Equation (17.1) as in the embedding technique.

For each column piece, discover the ordinary power and after that, the difference it and the typical power isolated from LSBs of the beginning 8 pixels in the related mapped RONI square. If they are not comparable then stamp the square as modified and displace the pixels in this piece with the isolated typical regard.

17.4   EXPLORATORY RESULTS

A MATLAB program for testing the execution of the proposed system was developed. For coordinating preliminaries, around 100 8-bit grayscale restorative images of different sizes and modalities such as CT scans, MRI yields, and Ultrasounds were used. Out of these 100 images, 35 restorative images are of CT scans, 40 remedial images are of MRI yield and 25 helpful images are of Ultrasounds. Zenith Signal to Noise Ratio (PSNR) and Weighted Peak Signal to Noise Ratio (WPSNR) [19] are used to measure the mutilation in the made watermarked helpful images.

Higher estimation of PSNR and WPSNR doles out less twisting in the water-marked image. The Mean Structural Similarity (MSSIM) document [20] metric is used to evaluate the closeness between the first and the watermarked remedial image. MSSIM regard is between −1 and 1. Regard 1 of MSSIM shows that the first and watermarked images are equivalent. Visual degradation in the watermarked image is assessed using the Total Perceptual Error (TPE) [21] metric. A lower estimation of TPE indicates less debasement in the watermarked image.

A segment of the therapeutic images used as a piece of our assessments are resized to 256×256 and understanding data of 0.5 KB measure is embedded inside ROI. A rectangular-shaped ROI is considered in each helpful image for duplicating the proposed procedure. There is no enormous visual complexity between the first, watermarked, and watermarked removed helpful images. The results gained in the wake of embedding the watermark into the three restorative images outlines the typical results received by watermarking the 100 remedial images used as a piece of our preliminaries.

A remedial image watermarking framework is effective if the PSNR estima-tion of watermarked and changed helpful images is more conspicuous than 40 dB [21]. In the proposed methodology, the PSNR and WPSNR estimations of water-marked and changed therapeutic images are more than 40 dB. The obvious change in the fundamental information of the watermarked helpful images is immaterial as the MSSIM regards all modalities of images are near 1. Also, the low ordinary TPE regards demonstrate less visual debasement in the watermarked therapeutic images.

The gate crashers are kept from getting information on ROI by scrambling it before introducing inside edge pixels. If an attacker perceives the ROI region and gets the LSBs of pixels inside ROI then he can't do anything with that data as is encoded by a riddle key. A segment of the top tier frameworks [9, 10, 11, 13] are not using any affirmation data, such as the hash estimation of ROI to check clearly whether the ROI is changed or not. Thusly, all pieces inside ROI must be checked reliably to recognize the proximity of changes. This checking system prompts time wastage when the watermarked remedial image isn't adjusted. Such time wastage isn't caused in the proposed system as it uses hash estimation of ROI to explicitly check whether the ROI is changed.

To test the execution of the proposed system to the extent of recognizing modi-fied squares inside ROI and recovering the original ROI, a modification inside ROI of the watermarked therapeutic images as mentioned was used. The proposed method recognized the modification inside ROI and recovered novel ROI. In a remedial image, the LSB of pixels inside RONI and edge are generally zero. Thus, the LSB of pixels inside RONI and edge are set to 0 in the wake of isolating intro-duced data from them.

For testing the capacity of the proposed system in perceiving adjustments at various regions inside ROI and recovering remarkable ROI, pixels at a number of territories inside the ROI of watermarked remedial images as mentioned were changed. The proposed methodology perceived all of the changes inside ROI and recovered one-of-a-kind ROI. A segment of the surveyed plans [10, 11, 13] can't recover ROI when modifications are started by aggressors at various zones inside ROI.

The proposed technique is created considering the doubt that the intruders all around undertake to change only the basic part, ROI, in the remedial images during their transmission. Thus, perceiving changes inside ROI and recovering remarkable ROI must be done before using the remedial image for finding and evading misdiagnosis. One of the hindrances of the proposed strategy is: the RONI and edge parts are not recovered accurately as LSBs of all pixels inside RONI and periphery are set to 0 consequent to removing embedded data from them. This imperative does not impact the efficiency of the proposed procedure as RONI and periphery parts of therapeutic images are not essential for choosing discovering decisions. Another obstacle is there is no security for the ROI recovery data that is embedded inside RONI. If the ROI is modified then the principal ROI can be recovered exactly when the RONI and edge of some portion of the watermarked therapeutic image are not struck by any confusion or not changed by intruders or not dealt with by normal image control exercises.

17.5   CONCLUSION

The proposed restorative image watermarking method conveys extraordinary watermarked remedial images. The watermarked helpful images look increasingly like one-of-a-kind remedial images such as PSNR and WPSNR estimations of watermarked therapeutic images are over 50 dB and MSSIM regards are over 0.93. The proposed system can be used with therapeutic images whose ROI part is up to 62% of the entire image. The proposed procedure uses only 8-bit affirmation and recovery data for each 4×4 square inside ROI. It recognizes any point of confinement changes inside ROI and recovers one-of-a-kind ROI. Exactly when the removed hash estimation of ROI matches with the recalculated hash estimation of ROI, by then the proposed technique doesn't check the squares inside ROI for distinguishing the proximity of changes. The computational multifaceted nature of the proposed strategy is less as it uses essential numerical estimations for making confirmation and recovery data, recognizing adjusted squares inside ROI and recovering one-of-a-kind ROI.

For future improvement, it is an endeavor to extend the system for helpful images whose pixels are addressed using 10-, 12-, or 16-bits and besides to keep up standard strikes, decline introducing mutilation inside ROI and recover the pixels inside ROI with their exceptional bits instead of with typical of pixels.

REFERENCES

1. G. Coatrieux et al., "Mixed reversible and RONI watermarking for medical image reliability protection", in *2007 29th Annual International Conference of the IEEE Engineering in Medicine and Biology Society*, pp. 5653–5656, IEEE, 2007
2. X. Luo et al., "A lossless data embedding scheme for medical images in application of e-Diagnosis", in *Proceedings of the 25th Annual International Conference of the IEEE Engineering in Medicine and Biology Society (IEEE Cat. No. 03CH37439)*, 2003, vol. 1, pp. 852–855, 2003.
3. J. J. Eggers et al., "Scalar SOSTA scheme for information hiding", *IEEE Transactions on Signal Processing*, 151(4), 2003, pp. 1003–1019.
4. N. A. Memon and S. A. M. Gilani, "NROI watermarking of medical images for content authentication", in *Proceedings of the 12th IEEE International Multitopic Conference, (INMIC'08)*, Karachi, Pakistan, 2008, pp. 106–110.
5. Chiang, Kuo-Hwa, et al. "Tamper detection and restoring system for medical images using wavelet-based reversible data embedding", *Journal of Digital Imaging*, 21, 2008, pp. 77–90.
6. A. Giakoumaki et al., "Multiple image watermarking applied to health information management", *IEEE Transactions on Information Technology in Biomedicine*, 10(4), 2006, pp. 722–732.
7. J. M. Zain and A. M. Fauzi, "Medical image watermarking with tamper detection and recovery", in *Proceedings of the 28th IEEE EMBS Annual International Conference*, 2006, p. 32703273.
8. J. H. K. Wu et al., "Tamper detection and recovery for medical images using near-lossless information hiding technique", *Journal of Digital Imaging*, 21(1), 2008, pp. 59–76.
9. K.-H. Chiang et al., "Tamper detection and restoring system for medical images using wavelet-based reversible data embedding", *Journal of Digital Imaging*, 21(1), 2008, pp. 77–90.
10. Liew, Siau-Chuin, and Jasni Mohamad Zain. "Reversible medical image watermarking for tamper detection and recovery", *2010 3rd International Conference on Computer Science and Information Technology*, Vol. 5. IEEE, 2010.
11. Baiying Lei et al., "Reversible watermarking scheme for medical image based on differential evolution", *Expert Systems with Applications*, 41(7), 2014, pp. 3178–3188.
12. Nisar Ahmed Memon et al., "Hybrid watermarking of medical images for ROI authentication and recovery", *International Journal of Computer Mathematics*, 88(10), 2011, pp. 2057–2071.
13. BW, Tjokorda Agung, and Febri Puguh Permana. "Medical image watermarking with tamper detection and recovery using reversible watermarking with LSB modification and run length encoding (RLE) compression", in *2012 IEEE International Conference on Communication, Networks and Satellite (ComNetSat)*, pp. 167–171. IEEE, 2012.
14. B. Nagaraj and P. Vijayakumar, "Evolutionary computation based controller tuning—A comparative approach", *International Journal Indian Pulp & Paper Technical Association*, 24(2), 2012, pp. 85–90.
15. O. M. Al-Qershi and B. E. Khoo, "ROI-based tamper detection and recovery for medical images using reversible watermarking technique", in *International Conference on Information Theory and Information Security (ICITIS)*, IEEE, 2010, pp. 151–155.
16. O. M. Al-Qershi and B. E. Khoo, "Authentication and data hiding using a hybrid ROI-based watermarking scheme for DICOM images", *Journal of Digital Imaging*, 24(1), 2011, pp. 114–125.
17. X. Deng et al., "Authentication and recovery of medical diagnostic image using dual reversible digital watermarking", *Journal of Nanoscience and Nanotechnology*, 13(3), 2013, pp. 2099–2107.

18. Baiying Lei et al., "Reversible watermarking scheme for medical image based on differential evolution", *Expert Systems with Applications*, 41(7), 2014, pp. 3178–3188.
19. Hussain Nyeem et al., "Ultilizing least significant bit-planes of RONI pixels for medical image watermarking", in *International Conference on Digital Image Computing: Techniques and Applications (DICTA)*, IEEE, 2013.
20. N. Ponomarenko et al., "Weighted MSE based metrics for characterization of visual quality of image denoising methods", in *Eighth International Workshop on Video Processing and Quality Metrics for Consumer Electronics – VPQM*, 2014.
21. W. Zhou, Atroni., "Image quality assessment: From error visibility to structural similarity", *IEEE Transactions on Image Processing*, 13(4), 2004, pp. 600–612.

18 Enhancement of Security and Fusion Techniques in Wireless Medical Sensor Network

K. Pranavi, D. Sathya, and D. Jagadeesan

CONTENTS

18.1 INTRODUCTION

In the past few years, wireless communication technologies and wireless medical sensor networks (WMSNs) have contributed more to the healthcare industry. Wireless sensor networks are a group of sensors distributed spatially to monitor and record the physical conditions of the environment. The captured data can be stored at a central location for further processing. The applications of WSN include military, healthcare, home, environmental, and other commercial areas. In wireless medical sensor networks, biosensors are used to collect personal medical data which contains sensitive information. WMSN paves the way for monitoring patients or adults remotely and provides access through wireless networks. The medical sensors are embedded in various medical instruments for disease detection, disease diagnosis, treatment, and management of illness. Hospitals, health insurance companies, government agencies, researchers, and patients can benefit from the rapid development of WMSN.

Though many advantages are available using WMSNs, providing security for healthcare data is a major concern. There are many existing systems that contribute to the protection of the privacy of data. A novel and lightweight secure system has been proposed to address the security challenges in WMSNs [1]. The design

DOI: 10.1201/9781003307716-18

was implemented in terms of protecting the privacy of data and providing backward secrecy. If the health records are digitalized, there is the possibility of attacks and misuse of the health data [2]. The solutions proposed were based on anonymity and authenticity, ensuring safeguards in emergency situations and privacy.

To achieve better security and privacy, asymmetric cryptographic techniques have been proposed [3]. The improved IBE-Lite cryptographic method implements more security provisioning than other conventional algorithms. Considering the resource efficiency factors, a novel key agreement scheme is employed for message authentication [4]. The inference results are effective in terms of security performance as well as power consumption analysis for Body Area Networks (BAN) The effectiveness of the system could be reliable not only by assuring security services but also enhancing QoS parameters. When the amount of data transmission is reduced, ultimately the energy consumption of the network will be reduced [5]. The biomedical sensors collect data periodically and combine them together to transmit to the network. Hence the lifetime of the network can be increased and the battery life can be optimized. Due to the data fusion approach, telemonitoring is achieved successfully to help the patients [6].

Although many cryptographic and data fusion approaches have been proposed to overcome security and QoS challenges, there are certain limitations. Since the lifetime of biosensors is less, they cannot be efficient for high-power sensor nodes. Also if the security requirements are not satisfied then the data could be attacked easily. To ensure security while maintaining data privacy and to enhance energy consumption, the proposed methods are helpful. With a detailed survey, various encryption methods such as lightweight algorithms, elliptic curve cryptography, IBE-Lite, and so on have contributed to providing security for health data. A survey on data fusion approaches could imply that energy consumption can be reduced and the network lifespan can be increased efficiently.

In this chapter, Section 18.1 reveals the outline of wireless medical sensor networks in healthcare and security issues. Both security and QoS parameters have been discussed in detail. Section 18.2 describes the related works accounting for the security and privacy issues as well as the energy efficiency mechanisms. Section 18.3 provides a comparative study of using Fuzzy logic in various environments for reducing energy consumption. Section 18.4 demonstrates the comparison of various cryptographic techniques and their advantages for security provisions. The conclusion and future work are explained in Section 18.5.

18.2   RELATED WORK

18.2.1   Health Data Security Using Cryptographic Techniques

In spite of the functionality, the issues of security and resource consumption in terms of battery, limited sensor nodes, and network lifespan are the important aspects to be considered while designing a network. In [1], the system used Advanced Encryption Standard (AES) and proxy-protected signatures by warrant to attain secure data transmission and data access control. As a result, the method provides privacy and backward secrecy. However, the system does not deal with energy consumption and is thus suitable only for low-power sensor nodes.

Privacy is a major issue for patients, as medical data contains sensitive information. The health record of a patient could be lost when it is digitalized. There are some cases in which attackers can acquire personal data. First, if the encryption or decryption is not fast enough, a DoS attack could take place. Second, the security requirements for the system must be satisfied. Third, if the sensors are smaller in size, then the sensor nodes may easily be lost. The authors proposed some technical solutions to achieve security while ensuring privacy. The system uses secret credentials, a pseudo random number generator and proof of knowledge [2]. The experimental result shows that access control, authentication, and confidentiality are obtained in an emergency situation.

IBE-Lite is a lightweight identity-based encryption for achieving security and privacy in body sensor networks [3]. A body sensor network (BSN) is a group of sensors used to collect personal medical data. IBE-Lite has been built using elliptic curve cryptography, which is a public key method suitable for BSN. The implementation took place through a proof-of-concept and the result provides reasonable performance in security and privacy protections for resource-constrained sensors. However, the system encounters a limitation that only 'n' secret keys can be released. If more than 'n' secret keys are released, the master secret key (X) is compromised to vulnerabilities. Conventional IBE does not have such inadequacy but it cannot be used on BSN hardware.

18.2.2 HEALTH DATA SECURITY USING BIOMETRIC TECHNIQUES

Besides the advantages of using BSN, there is a critical issue to be solved. Securing health data over a wireless environment with reduced energy consumption is a challenging task. The architecture allows neighboring nodes in BSNs to share a common key generated by electrocardiogram (ECG) signals [4]. The enhanced Jules–Sudan algorithm is employed to set up key management for the authenticating message. The experimental results could demonstrate the preservation of integrity and privacy is better than other conventional algorithms. The proposed ECG-IJS scheme achieves better security performance in terms of False Acceptance Rate (FAR), power consumption analysis, False Rejection Rate (FRR) and energy efficiency for BANs.

While ensuring security and privacy, developing an architecture is a major hurdle to be considered. The authors proposed a physical layer security algorithm based on behavior fingerprints using the wireless channel characteristics [5]. Compared to conventional authentication algorithms, it provides an effective and energy-friendly usability environment. The proposed system deploys three levels of security termed non-trust region, limited trust region, and trust region. As a result, no additional packages can be generated during the authentication process and this ensures battery life, a key factor of concern.

In an emergency situation, saving a patient's life is an important concern to be addressed. Due to huge population growth and insufficient care centers and staff, even mild health problems can become a major one. The security measure proposed has not only to sort out the problems but also to protect the privacy of personal data [6]. The proposed system contains a mobile emergency system (MES), which could serve in both the normal and emergency phases. To assure privacy and authentication, a secure mobile emergency system (SMES) has been deployed. Based on the alert, the response will be taken by the rescue team. By a mobile emergency call, the nearest authenticated rescue team will be available by which the safety of the patients can be assured.

In biomedical applications, resource constraints and security protections are major concerns. A traditional method provides integrity and privacy with respect to considerable memory and computational resources. Lightweight encryption and compressed sensing in wireless physical layer security are implemented [7]. The securing information is based on the use of a measurement matrix as an encryption key and assures security at the time of compressing the analog signals at the sampling time. The signals can be analyzed and stimulated using ECG. The data communication is reliable and secure and is much faster than AES.

Wireless medical transmissions of implantable medical devices (IMDs) should be secure against eavesdroppers and unauthorized users [8]. The proposed method uses full duplex secure communication of wireless IMD systems with a new protector. The protector is an externally worn jacket that provides physical layer security. The jacket is comprised of multiple sensors and is responsible for jamming extraction and Maximal Ratio Combining (MRC) reception. When the reception is performed, a spoofing-based beamforming technique called Multi-Input Multi-Output (MIMO) technology is used. The experimental results show that the protector is highly power efficient.

18.3 BIOSENSORS DATA FUSION

Moreover, security and privacy can be obtained in various implementations irrespective of the QoS parameters. One of the major factors is the reduction of energy consumption and increasing the network's lifespan. For instance, a method is introduced to fuse the collected data from the sensors before transmission to the wireless sensor network [9]. The authors proposed Fuzzy fusion logic to integrate the input data in each sensor and calculated the similarity percentage to identify the size of the packet to be sent. Based on membership functions of the inputs as defined in the database, a number of rules, and necessary interference, the similarity can be calculated. The simulation has been deployed in MATLAB and the result implies that energy consumption is reduced and network lifetime has been increased.

The fusion of collected data from sensors is not only helpful for QoS improvement but also helps in telemonitoring systems to ensure privacy. To safeguard independently living elderly people telemonitoring has been deployed. The developers proposed a multi-modal data fusion approach using fuzzy systems to integrate psychological, behavioral and acoustical environmental conditions [10]. The work has been implemented in the EMUTUM platform to address distress situations. The Fuzzy logic was used in various forms such as Fuzzy classifier to fuzzify the inputs, Fuzzy membership functions to categorize the levels of inputs and Fuzzy IF-THEN rules to defuzzify the inputs. As a result, the multi-modal data fusion approach increases the reliability of the whole system and ensures pervasive smart home health monitoring.

Recently, the contributions of wireless body sensor networks (WBSNs) are used widely in the healthcare industry. Biosensors gather personal medical data periodically and send it to the network where the data fusion occurs. Biosensors have a limited lifespan. In an emergency situation, processing massive amounts of data collected by biosensors and taking correct decisions are major challenging issues. The proposed method uses Fuzzy set theory for the data fusion approach and decision matrix to take the right decisions [11]. To optimize data transmission and energy consumption and to increase the lifespan of biosensors, the authors proposed a Modified Local Emergency Detection with an Adaptive Sampling Rate (Modified LED*) algorithm. The effectiveness of the proposed system is increased in data integrity and accurate decision making.

Cyber-Physical Systems (CPS) are widely used in healthcare where wireless sensor networks are the key elements. Computation, networking, and physical process are integrated to form CPS. The medical sensor generates many false alarms which reduce the system's effectiveness. To solve the issue, a Medical Fuzzy Alarm (MFA) filter is proposed using fuzzy IF-THEN rules. Fuzzy logic could handle unclear and inaccurate data [12]. The training of data sets and acquiring simulation results are done by means of the Weka tool, a Machine Learning technique. The experimental results show that the use of Fuzzy rules could achieve improved accuracy and increases the effectiveness of the system.

Internet of Things (IoT) devices have less computing power and energy. Hence the requirement of providing privacy to the data in e-healthcare is a major issue. Though many hard security solutions have been proposed, the authors intend to provide trust management methods by soft means. The proposed solution is analyzed in terms of RFID where more security services are primitive [13]. The presented solutions are appropriate for low-cost and commercial off-the-shelf (COTS) settings. A lightweight protocol for addressing privacy using ND-PEPS and extended metrics such as subjective logic is used as trust management methods. The contributions of this analysis develop cost-effective and non-invasive WMBANs for healthcare services.

Nowadays, health monitoring is widely used to address distress situations. The solution developed on Android-based mobile Data Acquisition (DAQ) is presented in [14]. The smart mobile device collects data from various wireless or wired sensors, analyzes it, and sends it for further processing. The proposed method uses cloud and smart information systems in which real-time sensors are used. Advanced RISC Machines (ARM7TDMI) are used as microcontrollers. RFID tags are attached to patients to identify and locate patients and to provide continuous monitoring information to health centers. The designed system provides low power consumption and portability.

Various wireless medical sensors collect biomedical parameters and combine them with Android-based smartphones to offer large functionality [15]. The physiological data is transferred through the Bluetooth HC-05 module to the smartphone. The ECG module using LM355 is used for signal processing. Hence the data are processed and sent to the healthcare centers. To respond at an emergency level, an alert system has been introduced by means of both SMS and Email alerts. As a result, the proposed system handles the emergency system effectively and is able to provide medical care as quickly as possible.

Recent studies have described how embedded and IoT devices are transforming rapidly in healthcare solutions. Wireless medical sensors and the IoT provide more convenience to patients as well as healthcare centers. Implementing complex anti-virus programs on IoT devices is costly and infeasible due to limitations in energy resources. A new framework called SYNDROME is introduced to monitor embedded medical devices externally [15]. The malware detector uses Electro-Magnetic (EM) signals to execute even in the absence of malware. Syringe pump, an embedded IoT device is used to evaluate SYNDROME. It is used to detect and stop the hijack.

18.4   COMPARISON OF SECURITY TECHNIQUES USED IN WMSN

The comparison of security techniques in WMSN with its advantages is listed in Table 18.1.

TABLE 18.1
Security Techniques

Application	Environment	Algorithm	Advantage
A Novel and Lightweight System to Secure Wireless Medical Sensor Networks [1]	E-healthcare	Advanced Encryption Standards(AES) and PSW technique	Secure data transmission and data access control system for MSNs
Privacy and Emergency Response in E-Healthcare Leveraging Wireless Body Sensor Networks [2]	E-healthcare	Anonymous credential, Pseudo random number generator, and proof-of-knowledge	Provides security and privacy in an emergency situation
IBE-Lite: A Lightweight Identity-Based Cryptography for Body Sensor Networks [3]	Healthcare monitoring	IBE-Lite	Balances security and privacy with accessibility for BSNs
ECG-Cryptography and Authentication in Body Area Networks [4]	Multimedia healthcare services	ECG-IJS scheme	Better security and achieve energy efficiency for BANs
Secure-Anonymous User Authentication Scheme for E-Healthcare Application Using Wireless Medical Sensor Networks [16]	E-healthcare	Secure-anonymous user authentication scheme (S-AUAS)	Provides high security along with incurring low computational and communication costs for WMSN

18.5   CONCLUSION

This chapter presented the security challenges faced by wireless sensor networks in healthcare and proposed various solutions to protect the privacy of health data. While addressing the battery life of sensor nodes, data fusion approaches can be used to reduce energy consumption. Data fusion techniques provide low-cost energy consumption with higher efficiency. By securing medical data, the sensitive information of an individual can also be secured. The various security techniques and data fusion mechanisms help in privacy protection. The analysis results could infer that data transmission occurring through WMSN needs to be secure as well as energy efficient. There is no such system to provide both security and energy efficiency in terms of high-power medical sensor nodes until now. Hence we have made a detailed analysis of encryption techniques and data filtration to provide privacy. Future work could take these issues into account and propose a secured system to transmit medical data and reduces energy consumption in WMSN.

REFERENCES

1. Daojing He, Sammy Chan, Shaohua Tang, "A Novel and Lightweight System to Secure Wireless Medical Sensor Networks", *IEEE Journal of Biomedical and Health Informatics*, 18(1), pp. 316–326, January 2014.
2. Jinyuan Sun, Yuguang Fang, Xiaoyan Zhu, Xidian University, "Privacy and Emergency Response in E-Healthcare Leveraging Wireless Body Sensor Networks", IEEE Wireless Communications, 17(1), pp. 66–73, February 2010.
3. Chiu C. Tan, Haodong Wang, Sheng Zhong, Qun Li, "IBE-Lite: A Lightweight Identity-Based Cryptography for Body Sensor Networks", *IEEE Engineering in Medicine and Biology Society*, 13(6), pp. 926–932, November 2009.
4. Zhaoyang Zhang, Athanasios Honggang Wang, V. Vasilakos, Hua Fang, "Ecg-Cryptography and Authentication in Body Area Networks", *IEEE Transactions on Information Technology in Biomedicine*, 16(6), pp. 1070–1078, November 2012.
5. Nan Zhao, Aifeng Ren, Zhiya Zhang Masood Ur Rehman, Xiaodong Yang, Fangming Hu, "Biometric Behavior Authentication Exploiting Propagation Characteristics of Wireless Channel", *IEEE Access*, 4, pp. 4789–4796, September 2016.
6. Shin-Yan Chiou, Zhen-Yuan Liao, "A Real-Time Automated and Privacy-Preserving Mobile Emergency-Medical-Service Network for Informing the Closest Rescuer to Rapidly Support Mobile-Emergency-Call Victims", *IEEE Access*, 6, pp. 35787–35800, June 2018.
7. Ruslan Dautov, Gill R. Tsouri, "Securing While Sampling in Wireless Body Area Networks with Application to Electrocardiography", *IEEE Journal of Biomedical and Health Informatics*, 20(1), pp. 135–142, October 2014.
8. Selman Kulaç, "A New Externally Worn Proxy-Based Protector for Non-secure Wireless Implantable Medical Devices: Security Jacket", *IEEE Access*, 7, pp. 55358–55366, April 2019.
9. Awat Mandeh, Keyhan Khamforoosh, Vafa Maihami, "Data Fusion in Wireless Sensor Networks Using Fuzzy Systems", *International Journal of Computer and Applications*, 125(12), pp. 31–36, September 2015.
10. Hamid Medjahed, Dan Istrate, Esigetel, Lrit, Jerome Boudy Jean-Louis Baldinger, Bernadette Dorizzi, "A Pervasive Multi-Sensor Data Fusion for Smart Home Healthcare Monitoring", *IEEE International Conference on Fuzzy Systems*, pp. 1466–1473, June 27–30, 2011.

11. Carol Habib, Abdallah Makhoul, Rony Darazi, Christian Salim, "Self-Adaptive Data, Collection and Fusion for Health Monitoring Based on Body Sensor Networks", *IEEE Transactions on Industrial Informatics*, 12(6), pp. 2342–2352, December 2016.
12. Wenjuan Li, Weizhi Meng, Chunhua Su, Lam For Kwok, "Towards False Alarm Reduction Using Fuzzy If-Then Rules for Medical Cyber Physical Systems", *IEEE Access*, 6, pp. 6530–6539, January 2018.
13. Denis TrcEk and Andrej Brodnik, "Hard and Soft Security Provisioning for Computationally Weak Pervasive Computing Systems In E-Health", *IEEE Wireless Communications*, 20(4), pp. 22–29, August 2013.
14. S. Lakshmanachari, C. Srihari, A. Sudhakar, P. Nalajala, "Design and Implementation of Cloud Based Patient Health Care Monitoring Systems Using IoT", *2017 International Conference on Energy, Communication, Data Analytics and Soft Computing (ICECDS)*, August 2017.
15. Nader Sehatbakhsh, Monjur Alam, Alireza Nazari, Alenka Zajic, Milos Prvulovic, "Syndrome: Spectral Analysis for Anomaly Detection on Medical IoT and Embedded Devices", *2018 IEEE International Symposium on Hardware Oriented Security and Trust (HOST)*, May 2018.
16. Yoney Kirsal Ever, "Secure-Anonymous User Authentication Scheme for e-Healthcare Application Using Wireless Medical Sensor Networks", *IEEE Systems Journal*, 13(1), pp. 456–467, September 2018.

19 An Exploratory Data Analysis on Cardiovascular Disease

V. P. Sumathi, R. Kalaiselvi,
A. Amritha, and R. Atshayasri

CONTENTS

19.1 INTRODUCTION

The World Health Organization (WHO) has estimated that 12 million deaths occur worldwide every year due to heart diseases. Half the deaths in the United States and other developed countries occur due to cardiovascular diseases. It is also the chief reason for deaths in numerous developing countries. Overall, it is regarded as the primary reason behind deaths in adults. The term heart disease encompasses the diverse diseases that affect the heart. Heart disease was the major cause of death in different countries including India. Heart disease kills one person every 34 seconds in the United States. Coronary heart disease, cardiomyopathy, and cardiovascular disease are some categories of heart diseases.

The term "cardiovascular disease" includes a wide range of conditions that affect the heart and the blood vessels, and the way blood is pumped and circulated through the body. Cardiovascular disease (CVD) results in several illness, disability, and death. The diagnosis of diseases is a vital and intricate job in medicine. Medical diagnosis is regarded as an important yet complicated task that needs to be executed

DOI: 10.1201/9781003307716-19

accurately and efficiently. The automation of this system would be extremely advantageous. Regrettably, all doctors do not possess expertise in every subspecialty and moreover, there is a resource shortage of persons in certain places. Therefore, an automatic medical diagnosis system would be exceedingly beneficial by bringing all of them together. Appropriate computer-based information and/or decision support systems can aid in achieving clinical tests at a reduced cost. Efficient and accurate implementation of automated systems needs a comparative study of various techniques available. This chapter aims to analyze different predictive/descriptive data mining techniques proposed in recent years for the diagnosis of heart disease.

The main functioning organ of the human body is the heart. Blood pressure, cholesterol, pulse rate, stress, and food habits are factors that contribute to heart disease. If the human heart isn't functioning it will affect the entire human body. Some risk factors of cardiopathy organ are family background, stress level, cholesterol level, age, diet, smoking, etc. The classification is done using Exploratory Data Analysis. Exploratory Data Analysis (EDA) is an approach to analyzing datasets that summarizes the main characteristics using visual methods. The purpose of using EDA is to uncover the underlying structure of a relatively large set of variables using visualizing techniques.

19.2   LITERATURE SURVEY

R. Dbritto and A. Srinivasaraghavan in their paper – Comparative Analysis of Accuracy on Heart Disease Prediction using Classification Methods – are concerned with four algorithms, Naïve Bayes, decision trees, k-nearest neighbor, and support vector machine. In this paper, the author suggests three phases. The size of the data set will be increased by using the first two algorithms and predicts heart disease by using logistic regression [1].

F. Lemke and J.-A. Mueller [2] proposed medical data analysis using self-organizing data mining technologies. In their paper, the authors constructed a decision tree by analyzing every medical procedure or medical problem. The problem for this paper was to collect a large amount of data for constructing the decision tree.

Seenivasagam [3] proposed the application of the EDA technique in healthcare and for the prediction of heart attacks. The author suggests four algorithms, rule-based, decision tree, Naïve Bayes and Artificial Neural Network for the massive volumes of healthcare data. The author uses the Tanagra data mining tool for their data analysis. The author analyzed 3000 instances with 14 different attributes.

L. Parthiban [4] presented a paper, on the coactive neuro-Fuzzy inference system (CANFIS) model for the prediction of heart disease. In this model the author compresses different techniques including the neural network adaptive capabilities, the Fuzzy logic qualitative approach and further integration with a genetic algorithm.

. Li, W., Han, J., & Pei, J [5] presented a paper – accurate and efficient classification based on multiple association rules. The authors proposed a novel method for their classification and named a predominant correlation. They proposed a fast filter method that can identify relevant features without pairwise correlation analysis.

Rani, K. U. [6] proposed a disease dataset using an EDA approach. In this paper, the author suggests an artificial neural network that combines forward and backward propagation algorithms. The experiment is conducted by single and multi-layered neural network models. Parallelism is implemented to speed up the learning process at each neuron in all hidden and output layers.

Dewan A [7] proposed the prediction of heart disease using EDA. In this paper, the author suggests a combination of data mining algorithms for predicting heart disease. The result of this paper concludes that the neural network is best among all the classification techniques for non-linear data. The author suggests that the back propagation algorithm is the best classifier of Artificial Neural Networks, which is a common method of training. But its main drawback is that it is held at a local minimum.

19.3   OBJECTIVE

1. To analyze the factors by using the Exploratory Data Analysis (EDA) technique.
2. To provide the graphical visualization of the statistical count on each category to predict the main risk factors and causes.

19.4   DATA ANALYSIS AND VISUALIZATION

19.4.1   DATA CLEANING AND DATA TRANSFORMATION

Derived Attribute:
   Age is represented in days. converting it into years. (age in days into age in years).
   Data Cleaning:

- If we look more closely at the height and weight columns, we will notice that the minimum height is 55 cm and the minimum weight is 10 kg.
- That has to be an error, since the minimum age is 10798 days, which equals 30 years.
- On the other hand, the maximum height is 250 cm and the highest weight is 200 kg, which might be irrelevant when generalizing data.
- We found this error while using the describe() function.

CARDIOVASCULAR DISEASE (CVD) DATA ANALYSIS AND VISUALIZATION

19.4.1.1   Diastolic bp and Systolic bp
- The normal range of diastolic bp is 60 to 80.
- The normal range of systolic bp is 90 to 120.

In this dataset, diastolic blood pressure is higher than systolic blood pressure in 1234 cases. So we are removing inaccurate data from the dataset. After removing the graph is plotted between diastolic and systolic bp.

In Figure 19.1, the X-axis represents the variable blood pressure. Blood pressure is divided into two types systolic blood pressure (ap_hi) and diastolic blood pressure (ap_lo). Systolic blood pressure refers to the pressure inside your arteries when

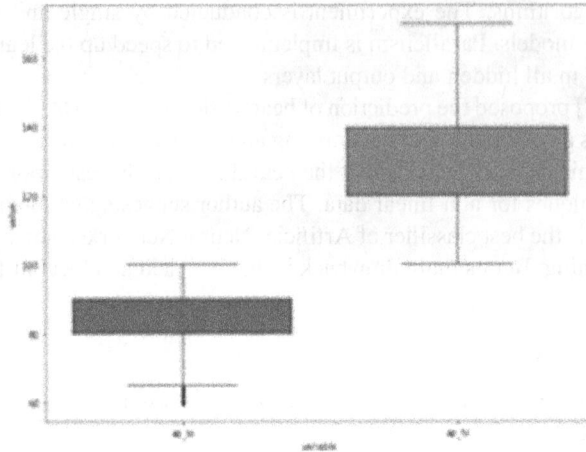

FIGURE 19.1   Systolic bp and diastolic bp

your heart is pumping; diastolic pressure is the pressure inside your arteries when your heart is resting between beats. The Y-axis represents the values. In most cases, diastolic bp lies in the range between 80 and 90 which is high when compared with a normal range and for systolic bp in most cases it lies above the normal range (90 to 120) which is between 120 and 140. Having a blood pressure range higher than the normal level would also be a cause of cardiac disease.

19.4.1.2   Correlation between the Attributes

There is a strong correlation between gender and height and cholesterol and glucose.

As we can see age and cholesterol have a significant impact, but are not very highly correlated with the target class.

Figure 19.2 shows the correlation map and that cholesterol and age have a powerful relationship with cardiovascular diseases. Number of persons affected by the disease was strongly correlated with age. Age playing vital role in the prediction of CVD

Figure 19.3's X-axis represents cardio which is a target variable. The Y-axis represents the number of persons affected or not affected by the cardio disease. About 33567 persons are not affected by the cardio which is represented by the first bar and the remaining 32626 persons are affected represented by the second dataset, both equally affected and not affected data are given.

19.4.1.3   Gender vs CVD

From Figure 19.4 we can infer the following:

   i. Women are mostly affected by cardiovascular disease in this taken dataset. Men are affected significantly less.

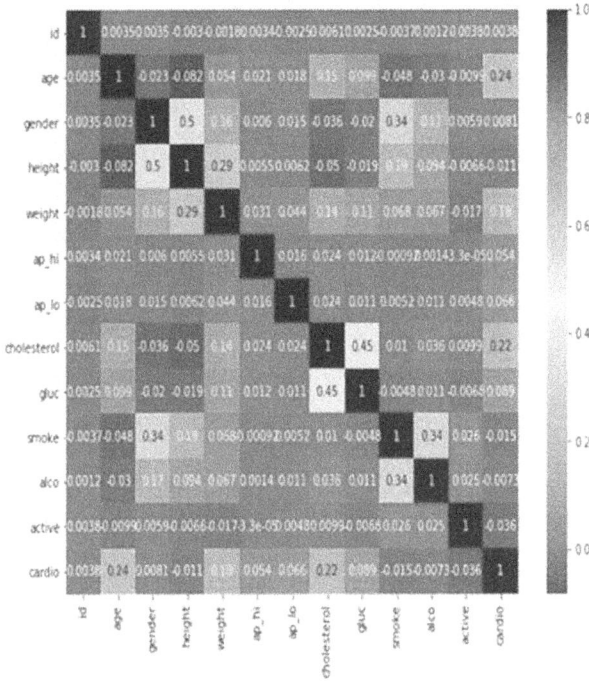

FIGURE 19.2 Correlation value between features

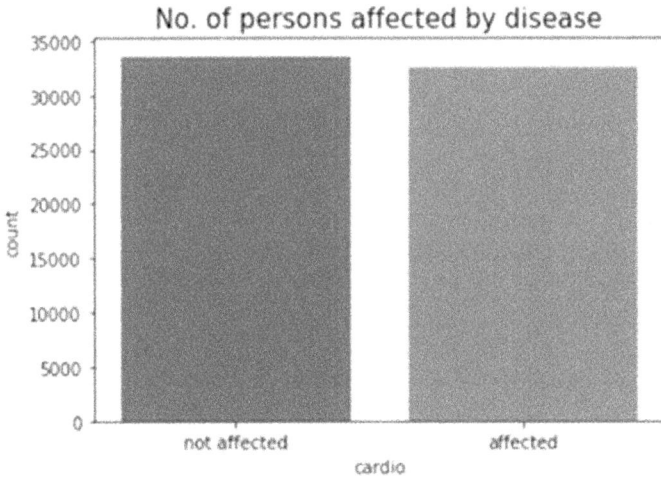

FIGURE 19.3 Disease affected

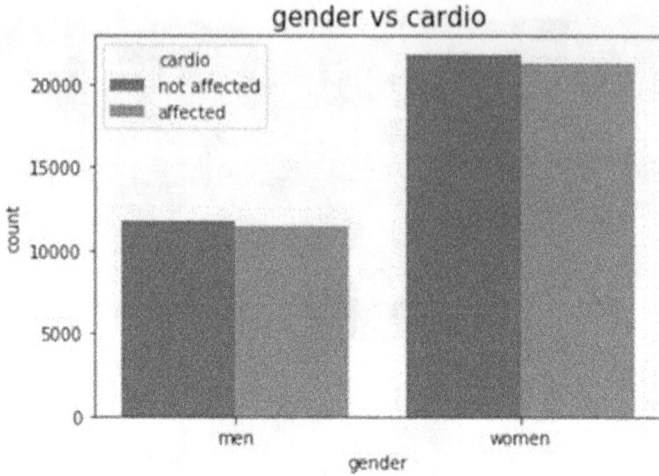

FIGURE 19.4 Gender vs cardio

  ii. After removing inaccurate data, the dataset contains 66193 observations. Among them, 11616 men are affected by the disease and 21210 women are affected by the disease.
 iii. Totals gender-wise, are women 42025 and men 23168.

19.4.2.3.1 Cholesterol vs CVD

Persons with normal cholesterol levels also affected by cardiovascular disease. But above normal and normal cholesterol were definitely affected by CVD. It can be clearly seen in the figure 19.5

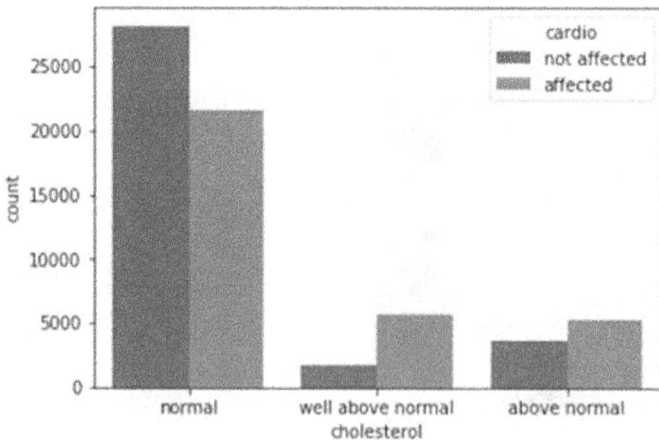

FIGURE 19.5 Cholesterol vs cardio

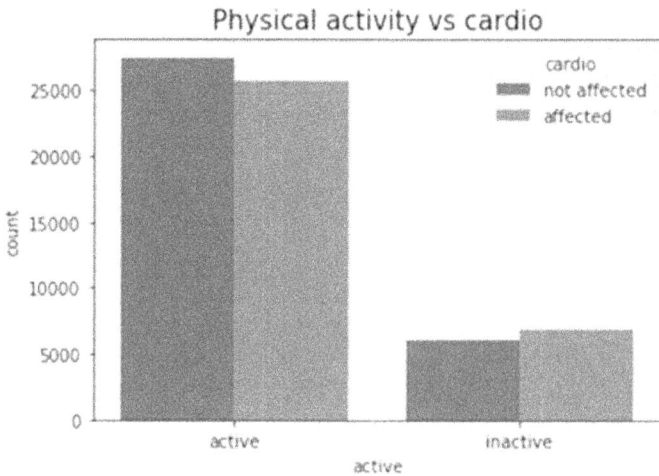

FIGURE 19.6   Activity vs cardio

19.4.2.3.2   Physical Activity vs Cardio

Persons who are inactive are affected by CVD. Persons who are inactive but maintain cholesterol and glucose levels are not affected by the disease. Though persons
doing physical activity are also affected because of other risk factors.

19.4.2.3.3   Glucose vs cardio

In Figure 19.7, the X-axis represents a glucose level that is normal, above normal,
and well above normal and the Y-axis represents the number of people. From Figure
19.7, It can be clearly seen that patients with CVD have higher glucose levels.

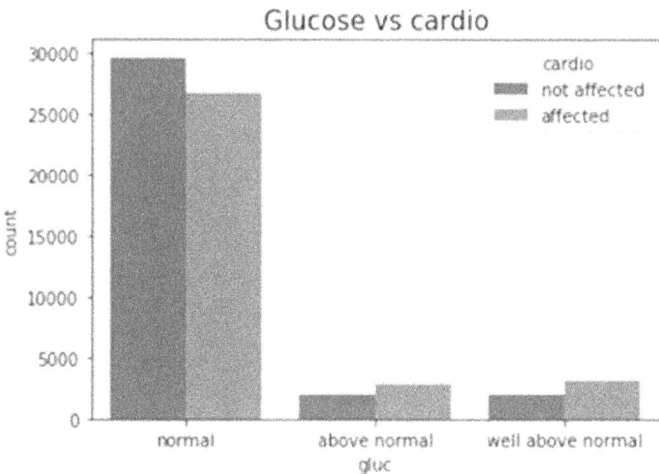

FIGURE 19.7   Glucose vs cardio

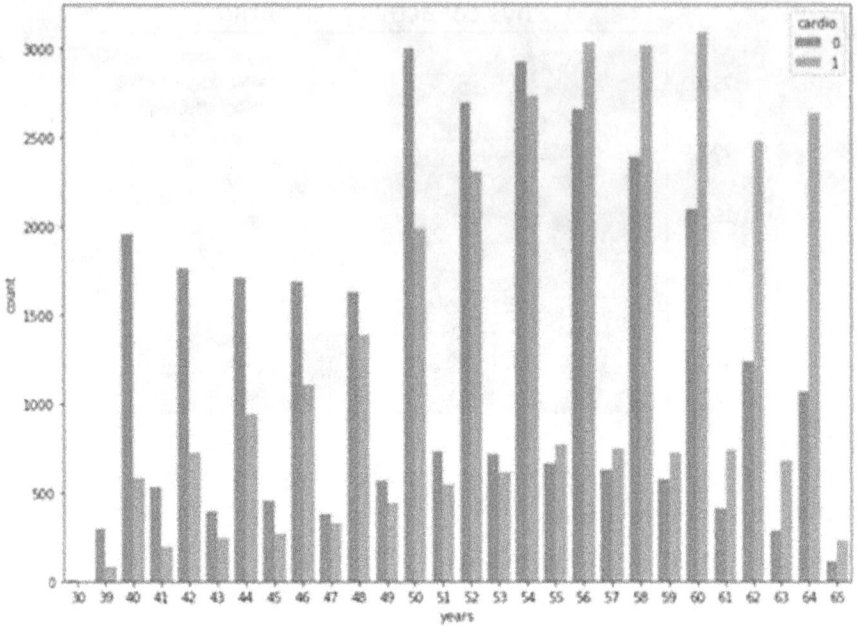

FIGURE 19.8 Age vs number of people affected

19.4.2.3.4 Analyzing the Age Attribute

In Figure 19.8, the X-axis represents years, that is the age of the person in years, and the Y-axis represents the number of people. The age is given in days in the dataset. we derived the attribute in years. From the graph, it is clear that with increasing age, the number of persons with cardio also increases.

Figure 19.9 shows that middle age people are the ones mostly affected by CVD.

- Young age – age between 30 and 40.
- Middle age – age between 40 and 60.
- Old age – age above 60.
- The maximum age in this dataset is 65.

In this dataset, age ranges from 30 to 65 years. In this, people between the ages of 40 and 60 are mostly affected. Above 60 is old age people who are also severely affected by CVD.

19.4.2.3.5 Gender vs Alcohol

In Figure 19.10, the X-axis represents gender, men and women, and the Y-axis represents the number of people consuming and not consuming alcohol. The graph shows

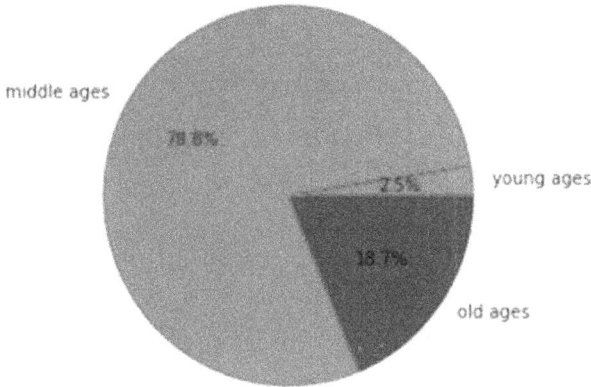

FIGURE 19.9   Category of age group affected

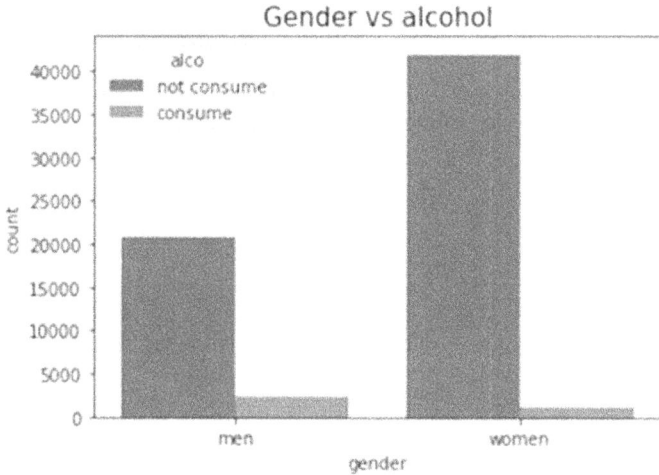

FIGURE 19.10   Gender vs alcohol

that in this dataset only less of patients consuming alcohol. This graph clearly shows that the number of men consuming alcohol is higher than women.

19.4.2.3.6   Gender vs BMI (Alcohol and CVD

- In Figure 19.11, X-axis represents gender, 1. women and 2. men and the Y-axis represents the persons BMI.
- BMI is calculated by dividing weight by height. The normal BMI range is between 18.5 and 25.
- Cardio 0 is not affected by CVD and cardio 1 is affected by CVD.
- Alcohol 0 represents not consuming alcohol and alcohol 1 represents consuming alcohol.

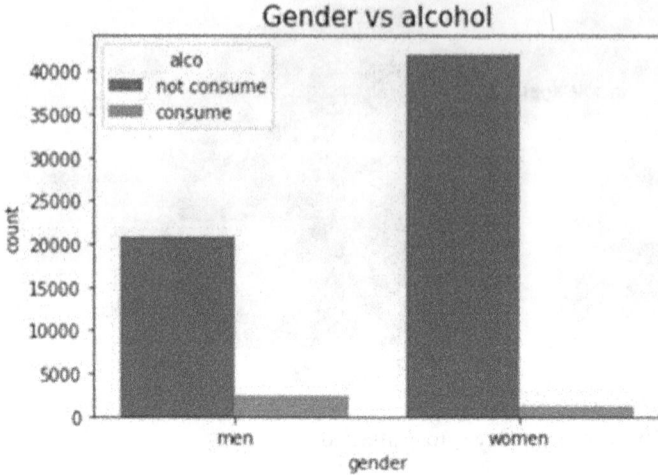

FIGURE 19.11   Gender vs BMI

- From the figure, we can conclude that women who consume alcohol are affected mostly by cardio disease.

19.5   CONCLUSION

The main purpose of the chapter is to analyze and predict the major risk factors. From a proper analysis of the dataset, different graphs were generated and visualized. From the graphs, we conclude that middle age and old age persons are the main victims of CVD. Women are the most affected persons in this dataset compared with men and mainly women who consume alcohol are affected mostly by cardio disease. Some of the major risk factors are cholesterol, glucose level, alcohol, and age.

19.6   FUTURE WORK

This work can be extended to higher levels in the future. The analysis can be referred for:

- A predictive model for *CVD* risk *prediction* that uses Machine Learning algorithms where the results from each graph of the paper can be taken as individual criteria for the Machine Learning algorithm.
- Processing resources for the data mining algorithm.
- Using the same dataset try identify important derived features that will be useful to predict women are affected by CVD

REFERENCES

1. R. Dbritto, V. Joseph. Comparative analysis of accuracy on heart disease prediction using classification methods, International Journal of Applied Information Systems (IJAIS) – ISSN: 2249-0868 Foundation of Computer Science FCS, New York, USA Volume 11 – No. 2, July 2016.
2. J. A. Mueller, F. Lemke. *Self-Organising Data Mining*, Libri, Hamburg, 2000.
3. Rajagopal Chitra, Dr. V. Seenivasagam. *Heart Disease Prediction System Using Supervised Learning Classifier*. SOCO, 2013. DOI: 10.9756/bijsesc.4336.
4. L. Parthiban, R. Subramanian. Intelligent heart disease prediction system using CANFIS and genetic algorithm. *International Journal of Biological, Biomedical and Medical Sciences*, 3(3), 278–281, 2008.
5. W. Li, J. Han, J. Pei. CMAR: Accurate and efficient classification based on multiple class-association rules. In *Proceedings 2001 IEEE international conference on data mining* (pp. 369–376). IEEE, 2001.
6. K. U. Rani. Analysis of heart diseases dataset using neural network approach. *arXiv preprint arXiv:1110.2626*, 2011.
7. A. Dewan, M. Sharma. Prediction of heart disease using a hybrid technique in data mining classification. In *2015 2nd International Conference on Computing for Sustainable Global Development (INDIACom)* (pp. 704–706). IEEE, 2015.

20 A Survey on Slight Barriers to Using Blockchain in Healthcare

V. Sudha and R. Kalaiselvi

CONTENTS

20.1 INTRODUCTION

Problems faced by users are different in kind and complexity which stimulates the advancements in technologies. Growing technologies led people to be able to appreciate almost all services with less cost. Cloud computing is one such horizontal innovation adopted by multiple fields and industries to make their functionality more efficient. All enterprises, including healthcare, are dealing with endlessly generated data. This forces industries to adopt cloud computing to face their computing and storage demands. In healthcare, it is required to maintain voluminous records such as different test results and medical prescriptions for each patient. Hence cloud computing is widely used in healthcare. Also, as most of the further decision making on diagnosis and treatment is primarily based on the medical records preserved, they must be available all the time. Hence, care must be taken to manage them. Not only does data storage in the cloud create security threats, but it also leads to loss of physical control and movement of data from one environment to another. To secure the data, standard security algorithms are applied. But this is not helpful in all stages. Thus, the existing challenge in healthcare is the data must be viewable by all but should not be modified once created. Blockchain can be used to achieve this. Blockchain technology is one of the most important and disruptive technologies in the world. Multiple industries are adopting blockchain technology to innovate the way they function. One industry that is looking to adopt blockchain is the healthcare industry. A copy of all the records will be available for all who are participating in the blockchain. There are several obstacles to adopting blockchain technology in healthcare data maintenance. Storing healthcare records using blockchain enables interoperability among doctors and reduces the time to retrieve the data, as all the

DOI: 10.1201/9781003307716-20

medical records with the diagnoses and symptoms are shared among all the participating nodes namely, doctors and researchers, this will lead to a reduction in the cost of drug discovery.

20.2 RELATED WORKS

In a lifetime, a patient may be visiting more than one physician. Hence, having a consolidated record or tracking their record is difficult. Also, different physicians may maintain the record in different formats. Hence, interoperability is very difficult. Nowadays some private agencies are assigned to maintain patient records. But accessing the data from it takes time. To overcome the above drawback, MedRec is proposed. In MedRec blockchain technology is used to keep track of the record. Bitcoin is used for transactions. Providing viewer permission is handled using a smart contract. Here, a patient can view the details of the viewer.

In this proposed method, after data is uploaded, viewer permission is issued only after getting consent from both patient and doctor. Three types of smart contracts are used for imposing the constraint. They are a registrar contract, a patient provider relationship contract (complete data about a patient is stored in more than one block, hence, usually a hash table is maintained to provide easy access) and a summary contract. Here, patients' privacy is maintained by encrypting the information using the patients' private key [1].

Blockchain can be used to increase interoperability while it helps in maintaining the privacy and security of data. As it contains inherent integrity it could conform to the strict adoption of legal regulations. Increased interoperability would be beneficial for health outcomes. Maintaining complete and correct medical records supports perfect diagnoses and treatment [2]. It is expected to record all the clinical trials completely with more care for accurate diagnoses. The probability of accuracy in Electronic Health Records (EHRs) could be less when compared to handwritten records due to typographical errors which are common in our fast-paced digital world [3]. Several secure web-based patient portals called *Patient Gateway* (PG) are available for producing more medication lists in the electronic health record (EHR). But accuracy of those patient portals is not significantly better when compared to handwritten. Medication lists represent one of the most important components of the EHR since they are used for filling refill requests, assessing quality, performing research, and informing computerized clinical decision support [4].

Though the use of electronic health records enhanced nursing work through increased information access, improved organization and efficiency, and helpful alert screens, it also involves hurdles. Nurses preferred to use electronic health records if they are comfortable with technology. Hence quality training on appropriate tools is required. From the study, it was observed that the use of electronic health records enabled the nurses to provide safer care but decreased the quality of care. Administrative implications include involving bedside nurses in system choice, streamlining processes, developing guidelines for consistent documentation quality and location, increasing system speed, choosing hardware that encourages bedside use and improving system information technology support [5].

There are various ways of collecting patient health records. The collected records may involve duplication or errors due to poor recording as many personnel are involved in the collection of clinical data such as healthcare experts, nurses, and lab technicians. Also, some people may not be able to express the data clearly due to their lack of proficiency [6].

In healthcare systems, the data are more hybrid and unstructured as they are collected from different equipment or systems connected with each other. Medical records do not contain the data collected not only from the prescriptions but also from other wearable devices.

Health Information systems consist of defining simple and extensible schemas for the most common and important types of data in healthcare and modeling a comprehensible, transparent and automatized ETL process (extract, transform, load) that would guarantee data integrity. Considering the sensitivity of the data in question, privacy is considered a crucial challenge in defining the model to ensure patient privacy is uncompromised [7]. Moreover, the healthcare systems are more patient-centric. Patients are involving themselves in the data.. Thus, all the clinical samples are taken from the patient [8].

A common dispute that was experienced in the fast-evolving technology is its unpopularity among its users. Though technology is going very fast in one side, the personnels in hospitals are still not aware of the technology and few are reluctant in adopting to the new technology [9]. It is important to monitor how physicians interact with EHRs to understand the their patient's health issues in order to fulfill the treatment requirements [10].

Though the patient's details are recorded correctly, it is not reliable when it is maintained only by healthcare centers. Hence data are stored in cloud storage where all the data maintenance is taken care of by Cloud Service Providers (CSP). Though the healthcare centers are free from data maintenance headaches, the patient does not have any control over their data when it is stored in the cloud. All medical records contain data such as patient identity, insurance details and much more [11]. However, storing health records exposes private information to hackers and leads to other security threats. Inadequate security measures of cloud storage lead to some gaps in its usage. In many cases, remote diagnosing of disease is needed with good scalability. The real-life constraints for achieving the same are security and privacy concerns, lack of trust, and lack of interoperable data standards. From this, it is inferred that we need a technology that is interoperable and more secure.

Though all the transactions are recorded, as standard medicine prescriptions followed in EHR, fake medicine verification can also be possible using blockchain. Moreover, as each node maintains a ledger, whenever there is a new transaction, a new block will be created and appended in all the local ledger copies. This leads to more time consumption. Hence, blockchain technology is costlier when applied to the application involving stream data. All the control of the transactions is done with the help of the program in a smart contract. The smart contract executes code on the blockchain network. This ensures data consistency. Blockchain must be able to scale up to meet the increase in the number of participating nodes. Decentralization, scalability, and consistency are the most three important properties of blockchain.

Among these three only two can be provided by the blockchain at a time. Always, there is a tradeoff between these [12].

According to HIPAA (Health Insurance and Portability Act) all the medical records with sensitive data redaction has to be provided for authorized users access. To facilitate data accessibility, data are stored in the cloud. Nowadays medical records are stored in the cloud to enable doctors to access the data wherever they are and also aid the patients to get service from doctors at distant places. Authentication and authorization are the two most important issues for the data in the virtual environment [13].

Though it is strongly believed that by moving healthcare data to the cloud better service can be provided, data security becomes a big issue. Communication security and confidentiality are the two needs of the users storing data in the cloud. Cloud computing fully depends upon the Internet. Hence, all security threats associated with the Internet are also applicable to cloud computing [14].

Hence, it is necessary to provide security and access control over this sensitive data [15]. All rules that govern the data set are done with the help of a smart contract. There is no ownership for a block maintained in the blockchain as it is a decentralized distributed database.

In a blockchain environment, all the nodes involved in the chain will have a copy of the transactions. This feature of blockchain makes it unsuitable for real-world problems. When the size of the blockchain increases then it is difficult for the nodes with small memory capacity to hold the entire blockchain in memory [16].

Different types of blockchain models are available. Some of them are Bitcoin, Ethereum, and Hyperledger. Among these three models, Bitcoin is the most mature model. Mining is the process of creating a block with transactions. The mining process must be very difficult to do and must be easy to verify. This is to avoid the intruder from creating false blocks in the chain. This is a reason for increasing the cost of block creation in the proposed technique. In the blockchain, a block can be modified. But due to this modification all the blocks after it has to be modified by recomputing the hash value. The cost involved in this modification is very high [17].

A false treatment can be given to a patient by creating a malicious blockchain and later, modifying the entire blockchain. Detection of malicious nodes in 3D space can be done by using the blockchain smart contract and the WSNs' quadrilateral measurement localization method, and the voting consensus results are recorded in the blockchain distributed [18].

Sometimes the blockchain can be affected by ransomware and blockchains are blocked until some amounts are paid [19]. An advanced biomedical/healthcare data ledger can be used for storing healthcare related data like diseases, symptoms, diagnostics, etc other than patient data can be stored in blockchain. As all the preserved records are in need of vast and frequent usage, only simultaneous read privileges for authorized users should be encouraged to avoid intentional changes to them. All these demands can be achieved by adopting blockchain technology [20].

One problem with the blockchain is its high transparency and low confidentiality. That is health-related information of a patient must be kept secret which is difficult to achieve in blockchain as everything is visible to all who participate in the blockchain activity. To overcome this problem usually instead of using the actual patient's name to store the data, pseudo names can be used. Thus, blockchain may lead to pseudonymity. To reduce this threat further a Virtual Private Network (VPN) can be used for healthcare data transmission [21].

20.3 CONCLUSION

This article provided an overview of the usage of blockchain technology in healthcare sectors. Though the advancement in the healthcare sector and technological innovation in electronic health record systems improves the availability of records, some issues that are faced due to blockchain technology are addressed. This can be avoided by combining secure record storage with granular access rules. Also, the framework proposes measures to ensure the system tackles the problem of data storage as it utilizes the off-chain storage mechanism of IPFS. Role-based access also benefits the system as the medical records are only available to trusted and related individuals.

REFERENCES

1. Ekblaw, A., Azaria, A., Halamka, J. D., & Lippman, A., "A case study for blockchain in healthcare: 'MedRec' prototype for electronic health records and medical research", white paper, August 2016.
2. Vazirani, A. A., O'Donoghue, O., Brindley, D., & Meinert, E., "Implementing blockchains for efficient health care: Systematic review", *Journal of Medical Internet Research*, 21(2), e12439, 2019.
3. Meinert, E., Alturkistani, A., Foley, K. A., Osama, T., Car, J., Majeed, A., ... & Brindley, D."Blockchain implementation in health care: Protocol for a systematic review", *JMIR research protocols, 8(2),* e10994, 2019.
4. Staroselsky, M., Volk, L. A., Tsurikova, R., Newmark, L. P., Lippincott, M., Litvak, I., Kittler, A., Wang, T., Wald, J., & Bates, D. W., "An effort to improve electronic health record medication list accuracy between visits: Patients' and physicians' response", *International Journal of Medical Informatics*, 77(3), 153–160, 2008.
5. Kossman, S. P., & Scheidenhelm, S. L., "Nurses' perceptions of the impact of electronic health records on work and patient outcomes", *CIN: Computers, Informatics, Nursing*, 26(2), 69–77, 2008.
6. Zhang, H., Yu, J., Tian, C., Zhao, P., Xu, G., & Lin, J. "Cloud storage for electronic health records based on secret sharing with verifiable reconstruction outsourcing", *IEEE Access*, 6, 40713–40722, 2018.
7. Koren, A., Jurčević, M., & Huljenić, D., "Requirements and challenges in integration of aggregated personal health data for inclusion into formal electronic health records (EHR)", *2019 2nd International Colloquium on Smart Grid Metrology (SMAGRIMET)* (pp. 1–5). 2019. doi: 10.23919/SMAGRIMET.2019.8720389.
8. Dubovitskaya, A., Xu, Z., Ryu, S., Schumacher, M., & Wang, F., "Secure and trustable electronic medical records sharing using blockchain", *AMIA Annual Symposium Proceedings*, 2017(2),650–656, 2017.
9. Gropper, A. "Powering the physician-patient relationship with HIE of one blockchain health IT", In *ONC/NIST Use of Blockchain for Healthcare and Research Workshop. Gaithersburg, Maryland, United States: ONC/NIST* 2016.

10. Asan, O., & Montague, E., "Physician interactions with electronic health records in primary care", *Health Systems*, 1(2), 96–103, 2012.
11. Kalaiselvi, R.,Vanitha, V., & Sumathi, V. P., "Protection of mental healthcare documents using sensitivity based encryption", *International Journal of Cloud Computing*, 10(1/2), 90–100, 2021.
12. El Sayed, A. I., Abdelaziz, M., Megahed, M. H., & Azeem, M. H. A., "A new supervision strategy based on blockchain for electronic health records", in *2020 12th International Conference on Electrical Engineering (ICEENG)* (pp. 151–156). IEEE, 2020.
13. Kuo, T.-T., Kim, H.-E., & Ohno-Machado, L., "Blockchain distributed ledger technologies for biomedical and health care applications", *Journal of the American Medical Informatics Association*, 24(6), 1211–1220, 2017.
14. Zhanf, P., White, J., Schmidt, D. C., Lenz, G., & Rosenbloom, S. T., "FHIRChain: Applying blockchain to securely and scalably share clinical data", *Computational and Structural Biotechnology Journal*, 16, 267–278, 2018.
15. Castaneda, C., Nalley, K., Mannion, C., Bhattacharyya, P., Blake, P., Pecora, A. et al., "Clinical decision support systems for improving diagnostic accuracy and achieving precision medicine", *Journal of Clinical Bioinformatics*, 5, 4, 2015.
16. Duan, J., Zhang, C., Gong, Y., Brown, S., & Li, Z., "A content-analysis based literature review in blockchain adoption within food supply chain", *International Journal of Environmental Research and Public Health*, 17(5), 1784, 2020.
17. Klinkmuller, C., Ponomarev, A., Tran, A. B., Weber, I., & Aalst, W. V. D. "Mining blockchain processes: Extracting process mining data from blockchain applications", in *International Conference on Business Process Management* (pp. 71–86). Springer, 2019.
18. She, W., Liu, Q., Tian, Z., Chen, J. S., Wang, B., & Liu, W., "Blockchain trust model for malicious node detection in wireless sensor networks", *IEEE Access*, 7, 38947–38956, 2019.
19. Gu, J., Sun, B., Du, X., Wang, J., Zhuang, Y., & Wang, Z., "Consortium blockchain-based malware detection in mobile devices", *IEEE Access*, 6, 12118–12128, 2018.
20. Partida, A., Criado, R., & Romance, M., "Identity and access management resilience against intentional risk for blockchain-based IOT platforms", *Electronics*, 10(4), 378, 2021.
21. Chen, J., "Hybrid blockchain and pseudonymous authentication for secure and trusted IoT networks", *ACM SIGBED Review*, 15(5), 22–28, 2018.

21 Encountering Privacy – Sensitive Information in Medical Documents

R. Kalaiselvi and V. P. Sumathi

CONTENTS

21.1 INTRODUCTION

With modernization, almost all fields depend on current day-to-day happenings, where it is required to share documents. Since medical documents contain confidential information such as patient name, disease diagnosis, age, address, etc., it is important to hide sensitive words to avoid one's personal dignity or life being damaged before sharing [1]. For instance, a celebrity who is a patient in a psychological treatment center may not like to reveal the disease details to anyone except the medical expert. But, different kinds of sensitive information could be found in the medical reports. Generally, sensitive data are categorized as mental health and abuse, sexually transmitted infections, and reproductive health.

As healthcare experts follow different styles or formats and terms while recording their diagnoses and suggestions/medications, the identification of sensitive terms becomes challenging. Also, the textual medical documents can be structured or unstructured. Though some tests are common and done for all patients, revealing the test result will cause damage to the patient's dignity. For instance, if a sensitive word or related symptom is present in a document, it should not be shared. This can be achieved by preventing access to parts of the document or the entire document [2]. Thus, preventing sensitive data from being shared with others.

DOI: 10.1201/9781003307716-21

In this chapter, a natural language processing technique is used for identifying phrases in clinical narratives which express a sensitive condition about the patient. The advantage of our technique is that it does not need any annotated or structured data. The problem of finding sensitive information in clinical notes is difficult as it is context-based. It applies an information-theoretic notion of term sensitivity to develop a sanitization method for textual medical data, which is well-suited to fulfill the privacy requirements stated by the current legislation on healthcare data protection. It helps in encountering semantic terms. Moreover, it exploits several general-purpose and healthcare (SNOMED-CT) knowledge bases to preserve data utility [3]. The proposed scheme is evaluated and compared against related works using documents describing highly sensitive medical concepts and realistic use cases based on existing regulations.

21.2 PROPOSED METHOD

The medical document is mostly a clinical narrative in the unstructured pattern which is given as input. Most sensitive elements related to patient confidentiality, such as names of diseases are in an irregular structure. Hence, there is a higher possibility of disclosure of sensitive information due to the usage of semantics (i.e., the fact that they refer to a specific and sensitive matter), rather than of their type. For example, diseases such as *flu* and *AIDS* in medical records may appear as plain words, but they are highly sensitive because of being discriminatory. *AIDS* has traditionally entailed social discrimination because of its transmission mechanism.

The goal of the proposed method is to automatically detect the sensitivity of a textual medical document according to certain privacy requirements, which would be usually specified by current legislation on medical data privacy as shown in fig. 21.1. The method consists of a preliminary step that takes the input privacy requirements and automatically sets the method parameters, these steps are executed after stemming the documents. This stemming is done in order to protect and identify the sensitive words from the document. Then the stemmed words are related to the input privacy requirements to identify the sensitivity of the document. The steps are explained in detail in the next subsections.

This preliminary step takes input from current legislation on medical data privacy specifying the kind of entities that should be protected because of their potential discrimination (e.g., AIDS/HIV, mental disease, sexually transmitted diseases, drug, or alcohol abuse). A database is used to compile a list of terms that will be considered sensitive. In this study, the sensitive medical terms outlined by an international non-profit standards development organization SNOMED-CT (Systemized Nomenclature of Medical Clinical Terms) are used as the database. Here, individual concepts are associated with lists of equivalent terms.

The list of sensitive terms will include the different synonyms and lexicalizations provided in SNOMED-CT for each entity to protect (e.g., for STDs: sexually transmitted disease, VD: venereal disease, etc.) and also all of its taxonomic specializations, which inherit the semantics of their sensitive ancestors (e.g., for STDs: gonorrhea, syphilis, chlamydia, etc.).We refer to the parts of the expression that are

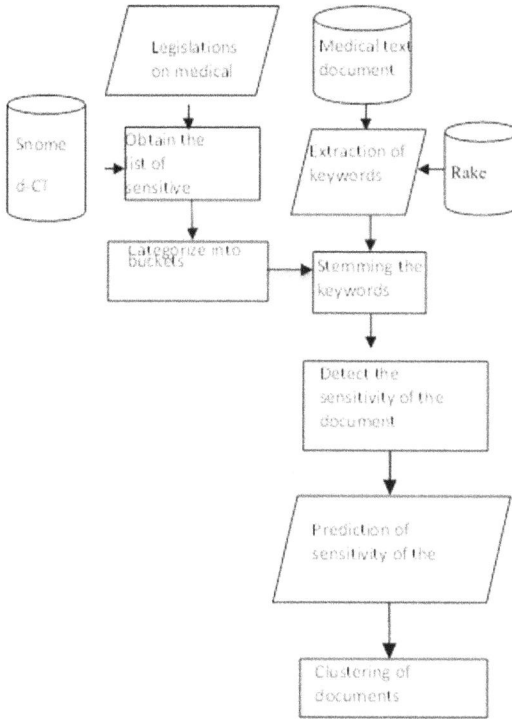

FIGURE 21.1 Working of the proposed method

considered sensitive by the privacy criterion. Official regulations such as HIPAA and States Safe Harbor Rules are made into account to maintain an individual's anonymity while preserving healthcare outcomes, which are useful for medical research [4]. Based on the level of sensitivity described by the medical legislature, the list of words obtained is categorized into three different buckets. This is done in order to categorize the sensitivity of the medical text document. These lists of items are stored in the database and are retrieved whenever needed.

21.3 EXTRACTING SENSITIVE WORDS FROM MEDICAL DOCUMENTS

Automatic identification of sensitive terms that best describe the subject of a document is a critical process without semantic analysis. The term is the terminology used for defining the words that represent the sensitive information contained in the document. Extraction of keywords is a vital task in text mining, information retrieval and natural language processing [4].

In order to extract the keywords from the medical text document a package called TEXTRACT is used. This package has interfaces such as a command line interface and a Python interface through that the keywords can be extracted from the content of all kinds of files, without any irrelevant markup. Most widely supported file formats include .csv via Python builtins, .doc via Antiword, .docx via Python docx2txt.

The keywords extracted using TEXTRACT are stemmed to get a word root. Stemming is the process of reducing the word to its stem i.e, its root form. For this stemming process, modules of the Natural Language Processing toolkit (NLTK) are used. It includes text analytics tools and tools for training data.

21.4   IDENTIFYING THE SENSITIVITY OF THE DOCUMENT

Based on HIPAA (Health Insurance Portability and Accountability Act) and SNOMED-CT rules, the extracted words of a document are classified into three categories, high-sensitivity keywords, medium-sensitivity keywords, and low-sensitivity keywords. Sample extracted words are shown in Table 21.1. Based on the ratio of sensitivity levels document is classified as a highly sensitive document or moderately sensitive document or low-sensitivity document.

Based on the level of sensitivity described by the medical legislature, the list of words obtained is categorized into three different buckets using the loop. Stemmed keywords from the medical text documents are compared with the buckets of medical terms from the database. When compared, word frequency in each bucket is calculated. Based on the frequency level in each bucket for a medical document.

In each document, three values – high, medium, and low – are calculated into three variables. If the combined sum of medium and low sensitive words is less than high sensitive words provided the value of high is less than a certain preset value then the document is classified as a highly sensitive document. If the value of low-sensitivity words is less than moderately sensitive words then it is classified as a moderately sensitive document. Otherwise, it is a low-sensitivity document and can be shared among others.

TABLE 21.1
Sample Words with Sensitivity Levels

Name	High	Medium	Low	Result
Sample 6	84	64	36	Moderate
Sample 5	31	30	33	Moderate
Sample 4	51	45	47	Moderate
Samples1	32	17	9	Moderate
Sample type2	19	13	8	Moderate
June	101	48	23	Very high
Pys-data	68	24	19	High
John	40	33	15	Moderate
Judy	22	20	15	Moderate
Assessment	88	69	30	Moderate
Sebastian	1191	1013	613	Moderate

21.5   CLUSTERING THE DOCUMENTS

Clustering is a process of grouping the items based on their similarities into clusters where an item in a cluster is more similar to the items within that group when compared to an item in another group. It is the process of segregating groups with similar traits and assigning them into clusters. Hard clustering groups each data item into one cluster completely whereas soft clustering assigns a probability of the item being in a cluster.

There are more than a hundred clustering algorithms known, but only a few of the algorithms are regularly used. Out of all the algorithms, we make use of k-means clustering which falls under centroid models. This kind of iterative algorithm focuses on identifying all the items which are closer in similarity with the centroid of the clusters. In these models, the number of clusters required at the end has to be mentioned beforehand, which makes it important to have prior knowledge of the dataset. Here the dataset is the result of the previous step. These models run iteratively to find the local optima [5].

K-means is simple, measurable, and easy to implement and also it is fast and more efficient in terms of computational cost, typically $O(K \times n \times d)$ than other clustering algorithms. It is computationally faster than other clustering algorithms if K is small when there is a large data collection [6, 7]. It also produces higher clusters than hierarchical clustering, especially if the clusters are globular. The clustering results can also be easily interpreted using this algorithm. It is also a great solution for pre-clustering, thus reducing the space into disjoint smaller sub-spaces where other clustering algorithms can be applied.

21.6   RESULTS AND DISCUSSION

Table 21.1 shows the words which are considered as low-, medium-, and high-sensitivity words. Based on the dictionary the words from the medical document are classified as high or moderate using a classification algorithm. Based on the results obtained clustering of the documents is done using a k-means algorithm to find the optimum solution. Further, the obtained result is verified with the results obtained from other clustering algorithms and the accuracy is determined.

21.7   RELATED WORKS

Many researchers have done research in extracting keywords from sensitive documents. Several researchers tried to automate the keyword extraction process. Dorr et al., have used the natural language processing technique with Machine Learning, a rule-based approach, and a keyword-based approach for tokening and gathering medical relevant terms from medical records [8]. The limitation of this approach is that it cannot distinguish medical acronyms.

Cumby et al., detected sensitive terms by combing natural language processing techniques with syntactic parsing of POS (Parts Of Speech) tagged tokens (to retrieve noun-phrase) [9]. As very rigid rules were used on web-query matching, this approach could not attain the goal fully. Thus, as the rules are employed on the estimation of sensitivity based on semantic matching for heterogeneous data, some sensitive words used in general were missed [10, 5].

Jindal et al., used the Metamap software and semi-supervised technique for finding phrases in clinical narratives which express a sensitive condition about the patient of the hierarchically structured data. The limitation of this approach is that it has a set expansion problem [11].

Ford et al., used the case detection algorithm and open source information extraction algorithms to get the information extracted from the text and to improve the accuracy of the case detection algorithm of the unstructured data. The limitation of this approach is that code-based document detection is very strict and needs supporting evidence [12].

Shu et al., used a Fuzzy fingerprint framework including packet collection, shingling, VirtualBox and DLD server to provide a privacy-preserving data-leak detection (DLD) solution to solve the issue where a special set of sensitive data digests is used in the detection of the Enron dataset [13]. The limitation of this approach is that the host-assisted mechanism for complete data-leak detection for large-scale organizations is difficult.

21.8   CONCLUSION AND FUTURE WORK

Thus the sensitivity of the medical text document can be identified using Machine Learning techniques and the documents can be clustered based on sensitivity using a classification algorithm. This method helps in identifying the sensitivity of the medical text automatically.

This work can be enhanced for multimedia documents such as speech and animated documents in the future.

REFERENCES

1. L. Deleger, K. Molnar, G. Savova, F. Xia, T. Lingren, Q. Li, K. Marsolo, A. Jegga, M. Kaiser, L. Stoutenborough and I. Solti, "Large-scale evaluation of automated clinical note de-identification and its impact on information extraction", *Journal of the American Medical Informatics Association JAMIA* 20(1): 84–94, 2013.
2. O. Ferrández, B.R. South, S. Shen, F.J. Friedlin, M.H. Samore and S.M. Meystre, "Evaluating current automatic de-identification methods with Veteran's health administration clinical documents", *BMC Medical Research Methodology* 12(109), 1–16, 2012.
3. O. Ferrández, B.R. South, S. Shen, F.J. Friedlin, M.H. Samore and S.M. Meystre, "BoB, a best-of-breed automated text de-identification system for VHA clinical documents", *Journal of American Medical Informatics Association JAMIA* 20: 77–83, 2013.
4. H. Harkema, J.N. Dowling, T. Thornblade and W.W. Chapman, "Context: An algorithm for determining negation, experiencer, and temporal status from clinical reports", *JBI* 42(5): 839–851, 2009.
5. P. Jindal, C.A. Gunter and D. Roth, "Detecting privacy-sensitive events in medical text", *Proceedings of the 5th ACM Conference on Bioinformatics, Computational Biology, and Health Informatics*, pp. 617–620, 2013

6. B. Divya and R. Kalaiselvi, "Review on confidentiality of the outsourced data research", *Journal of Science and Engineering Systems* 1: 1–7, 2017.

7. P. Jindal and D. Roth, "Learning from negative examples in set-expansion". *In*: *Proceedings of the 2011 Conference on International Conference on Data Mining (ICDM)*, IEEE Computer Society, 1110–1115, 2011.

8. D.A. Dorr, W.F. Phillips, S. Phansalkar, S.A. Sims and J.F. Hurdle, "Assessing the difficulty and time cost of de-identification in clinical narratives", *Methods of Information in Medicine* 45(3): 246–252, 2006.

9. C. Cumby and R. Ghan, "A machine learning based system for semi-automatically redacting documents". *In*: *Proceedings of the 23rd Innovative Applications of Artificial Intelligence Conference*, 1628–1635, 2011.

10. V.T. Chakaravarthy, H. Gupta, P. Roy and M. Mohania, "Efficient techniques for document sanitization". *In*: *Proceedings of the ACM Conference on Information and Knowledge Management*, CIKM 2008, 843–852, 2008.

11. E. Ford, J.A. Carroll, H.E. Smith, D. Scott and J.A. Cassell, "Extracting information from the text of electronic medical records to improve case detection – A systematic review", *Journal of the American Medical Informatics Association JAMIA* 23(5): 1007–1015, 2016.

12. X. Shu, D. Yao and E. Bertino, "Privacy-preserving detection of sensitive data exposure", *IEEE Transactions on Information Forensics and Security* 10(5): 1092–1103, 2015.

13. R. Kalaiselvi, V. Vanitha, V.P. Sumathi, "Protection of mental healthcare Documents using sensitivity based Encryption", *International Journal of Cloud Computing*, 10(1/2), 2021.

22 High Throughput Image Analysis to Assess the Morphological Characteristics of *Spirulina Platensis* Using ImageJ Software

Dinakari Sarangan, Lekha Kaushik,
Sowmiya Muthumanickam, and
Kandaswamy Kumaravel

CONTENTS

DOI: 10.1201/9781003307716-22

22.1 INTRODUCTION

In recent years the increase in resolution power and efficiency of microscopic image acquisition hardware leads to a tremendous growth of biological image datasets which has a lot of information. Extensively growing computerized image processing and analysis reduces visual inspection and manual measurement as they are potentially inaccurate and poorly reproducible. In the 21st century, many people adopt this computational process to accelerate repetitive tasks. ImageJ is a Java-based, open-source, image processing program developed by the national institutes of health and the laboratory for optical and computational instrumentation that provides extensibility via user-written Java plugins and recordable macros. ImageJ has both the combination of advanced image analysis and its simplicity to use makes it function effectively. It creates a platform between the expertise and the new users in their diverse environment for collaboration through ImageJ's mailing list. ImageJ can display, edit, analyze, process, save and print 8-bit color and grayscale, 16-bit integer, and 32-bit floating images. It can read many image file formats and can calculate area and pixel value statistics of user-defined selections and intensity-thresholded objects. The particle size and shape of talc powder was analyzed by ImageJ with customized tools for the segmentation of particles. In this chapter, we have analyzed the area perimeter solidity, circularity, and integrated density of *Spirulina platensis* which is widely used as a dietary supplement for vegan diets. The culture was cultivated in the laboratory and imaged with fluorescent microscopy. The above-mentioned parameters were analyzed using ImageJ and the results were recorded and presented in Figure 22.1.

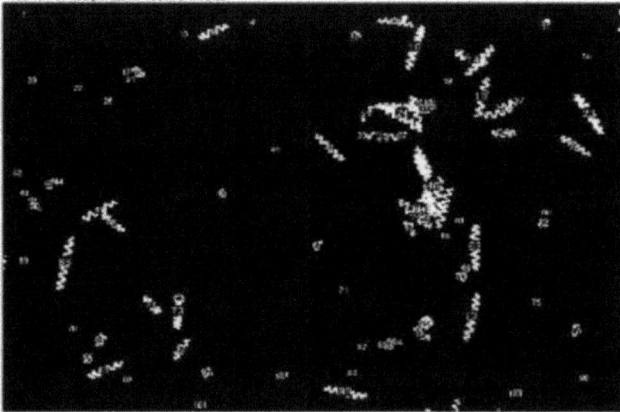

FIGURE 22.1 The image was taken when the outlines of trichomes were selected and displayed with numbers in ImageJ filtering once the information is obtained, filtering of impurities, and aggregates can be done by manual observation. This eliminates a significant number of objects which ImageJ assumes to be cells.

22.2 MATERIALS AND METHODS

22.2.1 CULTURE GROWTH

The spirulina culture was bought commercially from a spirulina farm. Since spirulina grows in an alkaline medium it is cultured in a Zarrouk's medium and the pH was maintained between 9–11. The culture is tested once a week with litmus paper. The flask was agitated 3–4 times a day clockwise and anticlockwise. The culture was exposed to alternate light and dark periods for photosynthesis.

22.2.2 SLIDE PREPARATION AND MICROSCOPY

The grown culture of two to three weeks is placed in a 1 ml centrifuge tube. It is centrifuged at 11000 rpm for 10 mins. The supernatant was discarded and the pellet was washed with 1 ml PBS buffer and resuspended in it. The above step was repeated three times and resuspended in 50 μl PBS buffer. The 20 μl culture was placed on a poly L-lysine slide, one drop of mounting solution was added, covered by a cover slip and sealed with nail polish. The slide was allowed to dry and viewed under a fluorescent microscope. The images were captured using an OLYMPUS CKX53 microscope attached with a 4× phase contrast objective lens of N.A. 0.10. The camera used was QImaging MicroPublisher 3.3 RTV. The pixel dimensions of all the images were 2048 × 1536. The images were saved in tiff format. The images were analyzed using ImageJ.

22.3 ESTIMATION OF GEOMETRIC PARAMETERS

22.3.1 SOLIDITY

$$Solidity = Area / Convex \ area.$$

Solidity gives us information on how convex or concave a shape is. A value of 1 means the shape is convex. The more it deviates from this value the less convex the shape is (Zdilla et al., 2016).

22.3.2 CIRCULARITY

$$Circularity = 4\pi \times [(Area)/(Perimeter)^2]$$

The value of circularity ranges from 0 to 1. A circularity of 1.0 means it is a perfect circle and any value less than 1 means it is less of a circle. The term circularity is used to compare unknown shapes with known shapes (Zdilla et al., 2016).

22.3.3 INTEGRATED DENSITY

Integrated density refers to the sum of all the pixels within the selected region, in this case, all the pixels within the trichome cell. This measure overcomes the issue of canceling out the brighter pixels by the dimmer ones as this is not an estimation of the average intensity of pixels.

22.4 IMAGEJ

ImageJ is an image analysis software that is Java-based and open-source. It was developed by the National Institutes of Health (NIH) image. It was developed by Wayne Rasband, NIH, Bethesda, Maryland, in 1997 and continues to grow today. ImageJ allows the measurement of various geometric parameters of cells. For example, perimeter, area, circularity, solidity, and integrated density can be measured using ImageJ. The particle size and shape analysis of talc powder was done previously. (Kumari and Rana, 2015). ImageJ can be downloaded by choosing the relevant platform, Mac, Windows, or Linux according to the individual's operating system. Once it has been downloaded it can be installed for free use. An image can be opened by choosing File→Open→Choose for the desired image. Another way is to drag the image to the ImageJ interface which will open it. The tiff images were converted into rgb images by Adobe Photoshop as ImageJ cannot analyze composite images. The rgb image is then converted into 8-bit by using the pathway image tab→type→8-bit.

22.4.1 INSERT SCALE AND SCALE BAR

The scale was set by using the line tool. A line was drawn for a known distance in the image on the box with a length of 200 μm. then using the analyze tab→set scale. A window will be displayed. The measurements were set as distance in pixels 234, known distance 200, pixel aspect ratio 1.047 and unit of length μm. The global option was clicked as the images follow the same scale. The scale was inserted by using the pathway analyze tab→tools →scale bar. The image was saved with a scale bar by using the pathway image→overlay→flatten→save.

22.4.2 IMAGE ANALYSIS

The threshold of the image was set by using the pathway image→adjust→threshold→apply. The threshold was adjusted until a maximum number of cells were selected. The measurements to be measured were set by using the pathway analyse→set measurements. The parameters which are to be measured were selected and okay was clicked. The outline was set by using the pathway analyze tab→analyze particles →outlines→okay. The image was displayed with outlines of trichomes selected with numbers. The result was obtained, summarized, and exported and saved as an Excel file (Schindelin et al. 2012, 2015).

22.5 RESULTS

The image was taken by using a fluorescent microscope with phase contrast N.A 0.10. The 4× magnification is used for imaging the trichomes. Twenty images were taken and saved in tiff format for analysis. In ImageJ the channel image was analyzed for certain parameters and the results were recorded and exported in an Excel file. The data were manually grouped into three categories based on the number of spirals. The alteration was made manually and only the trichomes data was included.

The error bars were calculated with mean and standard deviation (SD). A paired t-test was done and the P-value obtained was significant.

22.5.1   GROUPING OF CELLS

The trichomes were grouped into three categories such as s<5. the trichomes was sorted into these categories from the result. It was found that there were 201 trichomes in s<s5. All three categories were analyzed with the same number of cells to achieve even data in results. The mean and standard deviation for s<s5 were found to be 10.05 ± 2.90. All these categories were analyzed with 201 trichomes and the standard deviation was higher in s>5. The data were plotted in the graph seen in Figure 22.2.

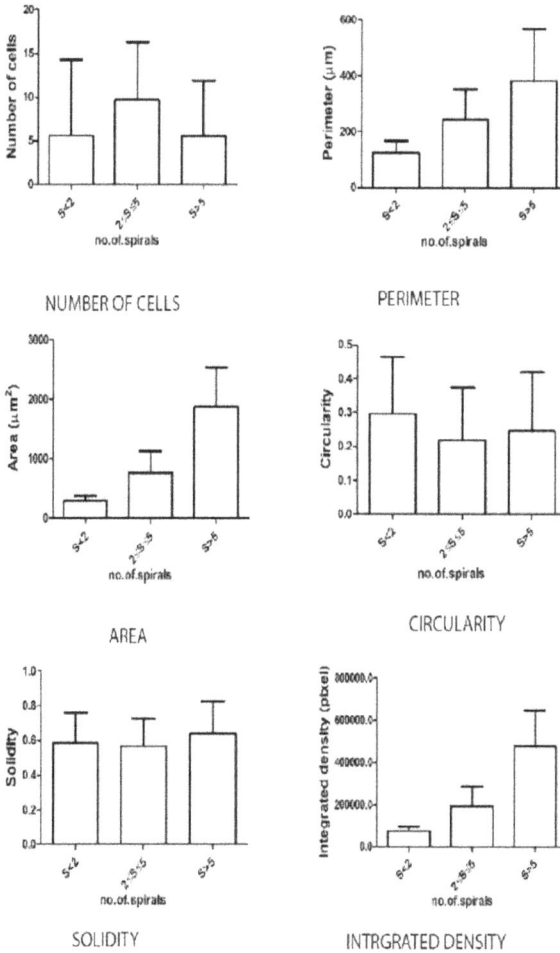

FIGURE 22.2   Bright field image of *Spirulina platensis* taken in Olympus CKX53 microscope. It is classified into three categories s<s5.

22.5.1.1 Analysis of Perimeter

The trichomes categorized into three categories were analyzed for perimeter in ImageJ. The data were found for 201 cells in each category and plotted as the graph seen in Figure 22.2. The 201 trichomes were selected on the basis of the perimeter in each category. It was found that the mean and standard deviation for s<s5 were found to be 573.18 ± 52.74. From this data, it was found that the mean and standard deviation is higher in the category of trichomes s>5.

22.5.1.2 Analysis of Area

The area was measured for 201 trichomes in each category and the results were recorded as seen in Figure 22.2. The mean and standard deviation for s<s5 were which has a range of 1868.02 ± 679.12. The third category is >5 is higher in the area as it has spirals greater than 5.

22.5.1.3 Analysis of Circularity

The circularity of the 201 trichomes was analyzed in ImageJ and plotted as the graph seen in Figure 22.2. The mean and standard deviation for the three categories was found to be s<s5 was found to be 0.26 ± 0.17. The circularity will always be less than 1. The results interpret that trichomes with spirals of 2<s<s5 which has a range of 0.64 ± 1.94. The spirals s>5 have a higher mean and standard deviation.

22.5.1.4 Analysis of Solidity

The solidity of the trichomes was measured and the data was plotted in the graph see in Figure 22.1. Three categories were taken under consideration with 201 trichomes respectively. In the results, it was found that the mean and standard deviation for spirals s<s5 which has a range of 0.64 ± 1.94. The spirals s>5 have a higher mean and standard deviation.

22.5.1.5 Analysis of Integrated Density

The integrated density of the trichomes was measured and the results were analyzed for 201 trichomes under three categories and plotted as the graph shown in Figure 22.2. The mean and standard deviation for the three categories are s<s5 which has a range of 476346.8 ± 173177.6. The mean is found to be higher in the third category s>5.

22.5.1.6 Calculation of the Error Percentage

The error percentage was calculated under three categories s<2, 2<s5 for parameters such as area, perimeter, solidity, circularity, and integrated density. The total error percentage was calculated by, (number of cells counted by ImageJ – number of cells left by ImageJ) / total number of cells present in the image. The error percentage was found to be 31.5%.

22.6   DISCUSSION

In ImageJ the parameters such as area, perimeter, solidity, circularity, and integrated density were analyzed. Since it is a semi-automated tool the data were recorded but not grouped under certain categories. The three categories were assigned manually according to the number of spirals that is s<2,2<s2. The same number of trichomes was taken in order to analyze the three parameters with 201 cells. ImageJ is excellent at finding the above-mentioned parameters for irregular shapes. They are easy to use, cost-efficient, and the results are displayed rapidly. Integrated density has been opted for as it would be a direct measure of all the pixels enclosed in the selected area. The error percentage indicates that ImageJ works fairly on irregular particles but is not accurate. Another drawback is it is semi-automated and it needs a manual worker. It does not classify the cells under a certain threshold it measures the parameters for all the cells present in the image. So to minimize the error we had taken an even number of cells in each category. It is separated on the basis of the perimeter. For larger cells and cells under a given threshold, it is efficient to use.

REFERENCES

Kumari, R. and N. Rana (2015). "Particle size and shape analysis using ImageJ with customized tools for segmentation of particles." *International Journal of Engineering Research* 4(11): 23–28.

Schindelin, J., et al. (2012). "Fiji: An opensource platform for biological-image analysis." *Nature Methods* 9(7): 676.

Schindelin, J., et al. (2015). "The ImageJ ecosystem: An open platform for biomedical image analysis." *Molecular Reproduction and Development* 82(7–8): 518–529.

Zdilla, M. J., et al. (2016). "Circularity, solidity, axes of a best fit ellipse, aspect ratio, and roundness of the foramen ovale: A morphometric analysis with neurosurgical considerations." *The Journal of Craniofacial Surgery* 27(1): 222.

23 Fluorescent Probes to Visualize Subcellular Localization of Proteins in Bacterial Cells

Yogesan Meganathan,
Deenadayalan Karaiyagowder Govindarajan,
Muthusaravanan Sivaramakrishnan,
and Kandaswamy Kumaravel

CONTENTS

23.1 INTRODUCTION

The visualization and trafficking of proteins in their native environment show detailed information on the function of the proteins. In bacteria, protein localization is initially studied using immunostaining (Coons, Creech and Jones, 1941). Later protein tags and fluorescent proteins were developed. Protein tags are short stretches of peptides that are

DOI: 10.1201/9781003307716-23

genetically engineered to the protein of interest (POI) which helps in the localization and purification of protein. Fluorescent proteins are proteins that produce fluorescence when expressed in living cells. These help in visualizing the protein present inside the cell using fluorescent microscopy (FM), transmission electron microscopy (TEM) and super-resolution microscopy (SRM). Live-cell imaging to track proteins in the subcellular region at different timeframes shows the origin and translocation of the protein to the site of action. In *Enterococcus faecalis*, Sortase A (SrtA), an integral membrane protein plays a major role in the assembly of virulence factors, to study the localization of the protein, Human influenza (HA) was tagged to the SrtA (Kandaswamy et al., 2013), and their discrete sites were localized. During cell division in bacteria, the synthesis of cell wall peptidoglycan is a curial part of cytokinesis, which was tracked by a self-labeling enzyme tag, Halo tag, and NeoGreen fluorescent protein. The two different proteins FtsA and FtsZ were tagged to the two different tags, and their mechanism of action was predicted (Bisson-Filho et al., 2017). DNA integrity scanning protein (DisA), a protein that scans for damages in DNA before replication delays the process of replication when damages are found. The localization pattern of the DisA protein to DNA was studied using the GFP tag (Bejerano-Sagie et al., 2006). Another protein called MreB, (an actin homolog) controls the length of the cell wall in *Escherichia coli* (*E. coli*) and is present on the cell surface that guides bacterial cell elongation. Tagging a GFP molecule to the protein elucidates the assembly of the MreB protein on the cell surface (Bendezú, Hale, Bernhardt and De Boer, 2009). Subcellular proteins fused to fluorescent tags are imaged by various biophysical methods such as wide-field FM, fluorescence recovery after photobleaching (FRAP) and fluorescence resonance energy transfer (FRET). FM techniques rely on the absorption of light at one wavelength (excitation) and emission at a longer wavelength. FRET describes the energy transfer mechanism between the chromophores (Sekar and Periasamy, 2003). A donor molecule (excitation state) may emit its energy to the acceptor molecule. By the dipole–dipole coupling mechanism, a non-radiatively transfer of energy takes place between the molecules. Energy transfer efficiency is inversely proportional to 10^6 times the distance between two molecules. Because of this FRET is more sensitive to small changes in the distance (Fairclough and Cantor, 1978). Whereas FRAP determines the kinetics of diffusion using fluorescent molecules. When a part of the fluorescent image is bleached, the fluorescent molecule recovers its fluorescence over time by the unbleached fluorescent in that environment. The BRET method of predicting PPIs relies on the mechanism of FRET (Bacart, Corbel, Jockers, Bach and Couturier, 2008; Dragulescu-Andrasi, Chan, De, Massoud and Gambhir, 2011). It has luciferase and fluorescent molecules like GFP and YFP. A split luciferase complementation system is mainly used for the study of PPIs (Paulmurugan, Umezawa and Gambhir, 2002). These methods were initially developed in eukaryotic systems and now it has also been employed on prokaryotes mainly in bacteria. In this chapter, we critically evaluate the merits and demerits of using fluorescent proteins (FPs), protein tags, and self-labeling enzyme tags.

23.2   PROTEIN TAGS

The short stretch of protein sequences that are attached to the protein of interest (POI) helps the purification and identification of that protein. Generally, protein tags can be categorized into four types. They are affinity tags, epitope tags, fluorescent tags, and self-labeling enzyme tags. Affinity tags such as poly His, poly Arg, His tag, and GST have an affinity toward a compound that helps in the purification of proteins, particularly recombinant proteins. Tags help in the easy purification of protein from the crude mixture using column chromatographic techniques. Epitope tags are short peptide sequences that are immunogenic (Munro and Pelham, 1984). These tags use the antibody molecule linked to an enzyme or substrate or radioactive or fluorescent molecule. Most of the focal localization techniques use epitope tags.

Generally, protein tags were used for the purification of proteins but they can also be used in the focalization of subcellular proteins. To explore the mechanism of biofilm formation by *Vibrio cholerae*, a Gram-negative bacterium, three matrix proteins, RbmA, RbmC, and Bap1, that play an important role in biofilm formation were tagged to three different tags Myc, FLAG, and HA tag, respectively. By supplying different cyanine-labeled primary antibodies to the growth medium, proteins were labeled and termed immunostaining. Live-cell imaging of *V. cholerae* cell from a single founder cell to biofilm formation pictured the space and time relation of these proteins. RbmA was first observed in the cell at the discrete site of the cell surface. Bap1 was found to be secreted at different distances from the founder cell. RbmC appeared at the last and was found to form an envelope attached to Bap1 (Berk et al., 2012).

Similarly, the biofilm formation of Gram-positive bacteria, *Enterococcus faecalis* was explored with a HA tag. Based on the immunogold microscopy method, *E. faecalis* was grown in the secondary phase and HA probed for sortaseA (SrtA) revealed the localization of SrtA at single loci (Kandaswamy et al., 2013; Kline et al., 2009). Although PTs are highly favorable for focal localization and purification it has their limitations such as size, molecular weight, and the maturation of protein. Some of the commonly used tags and their information are listed in Table 23.1.

23.2.1   Self-Labeling Enzyme Tags

Synthetic ligands coupled to markers such as fluorescent are catalyzed by enzymes under physiological conditions. Such enzymes are called self-labeling enzymes. Attachment of these synthetic ligands to the enzyme by a covalent bond was highly specific and irreversible (Liss et al., 2015). Such enzymes are tagged to POIs and synthetic ligands are provided exogenously. SNAP (Liss et al., 2015; Schermelleh, Heintzmann and Leonhardt, 2010), CLIP (Gautier et al., 2008) and HALO (Los et al., 2008) are the most commonly used self-labeling enzyme tags. SNAP-tag is derived from a human DNA repair protein, O^6alkylguanine-DNA alkyltransferase (hAGT), that irreversibly transfers the alkyl group from its substrate (O^6alkylguanine-DNA) to one of its cysteine residues. hAGT readily reacts with the O^6-benzylguanine derivatives (BG) which are used as a synthetic ligand that are coupled to fluorescein or other markers (Gautier et al., 2008). In a study on G-protein coupled receptor (GPCR), the N-terminal of the $GABA_B$ (a member of the GPCR family) was tagged to the SNAP-tag. The BG derivative has attached to europium cryptate or acceptor. It reveals the dimerization of $GABA_{B1}$ and $GABA_{B2}$ subunits at the cell surface (Maurel et al., 2008).

TABLE 23.1
Commonly Used Protein and Enzyme Tags

Tag	Residues	Size (Amino acid)	Molecular weight (KDa)	Pros	Cons	Reference
Human influenza (HA)	YPYDVPDYA	9	1.12	Short in size, rarely affect the heterologous protein expression.	Inefficient for the purification of native proteins for structure and function studies.	(Berk et al., 2012; Kandaswamy et al., 2013)
Myc epitope	EQKLISEEDL	10	1.2	Smaller in size doesn't interfere with the protein of interest.	Interfere with translocation into the secretory pathway hence it is advisable to avoid tagging to secretory signal peptides.	(Berk et al., 2012)
Strep	WSHPQFEK	8	1.06	High mechanical stability compared to the strongest non-covalent linkages was currently available.	Strep-tag can only be fused to the C-terminal end of recombinant proteins.	(Ke, Landgraf, Paulsson & Berkmen, 2016)
Flag	DYKDDDDK	8	1.01	More hydrophobic compared to other epitope tags and therefore less likely to denature protein of interest.	The tyrosine residue can be sulfated, which can affect antibody recognition of the FLAG epitope.	(Berk et al., 2012)
Self-labeling enzyme tags						
Halo	Haloalkane dehalogenases	297	33	Relatively small and there is no interference by endogenous mammalian metabolic reactions.	The slow kinetics of substrate binding of HaloTag posed a significant limitation.	(Los et al., 2008)
SNAP	O^6-alkylguanine-DNA alkyltransferase	182	19.4	It can be used for the measurement of protein half-lives in vivo and small molecule-protein interactions.	Because of its low reactivity, it is advised not to use it for the study of pulse-chase experiments.	(Shimomura, Johnson & Saiga, 1962)
CLIP	O^2-alkylcytosine-DNA-alkyltransferase	182	19.4	High specificity, small tag sizes.	Reaction time is higher and requires a high concentration of substrate.	(Gautier et al., 2008)

CLIP tag derived from O^2-alkylcytosine-DNAalkyltransferase that acts on O^2-benzylcytosine derivatives. Whereas HALO tags are modified bacterial haloalkane dehalogenases which remove halides from the aliphatic hydrocarbons by nucleophilic displacement mechanism (Janssen, 2004; Los et al., 2008). A ligand-bound labels tetra-methyl rhodamine (TMR), a red fluorescent molecule that is non-toxic, monomeric and membrane-permeable (Perkovic et al., 2014). Compared to other PTs, self-labeling enzyme tags enable the investigation of fast biological processes, linkage of the fluorescent to the ligand is covalent thus conferring stability of labeling and the choice of labeling groups enables the selection of low toxic fluorescent molecules. Because of the use of enzymes, these tags are highly specific. Enzymes currently used in the studies use higher substrate concentrations and longer reaction times are required and labeling of intracellular protein remains challenging (Lotze, Reinhardt, Seitz and Beck-Sickinger, 2016).

23.2.2 Fluorescent Protein Tags

The era of using green fluorescent protein started in the early 1960s after it was discovered during the study on the bioluminescence of *Aequorea Victoria* (Doi et al., 2002; Ormö et al., 1996). Wild-type GFP (WT-GFP) is a 25kDa protein that has a stable chain of 238aa and forms up 11 strands β-barrel structure wrapped around a single α-helix. The crystallographic structure of GFP elucidates 42Å long and 24Å diameter barrel structures (Figure 23.1). The fluorescent part of GFP is the p-hydroxybenzylidene-imidazolidinone (chromophore). The sidechain of the serine at 65th position and glycine at 67th form the imidazolidinone ring by elimination of water and further oxidation of tyrosine residue at double bond thereby generating a complete chromophore with excitation at 395nm and subsequent emission at 508nm. A combination of Ser-Tyr-Gly was present in many other proteins but not all others are fluorescent (Follenius-Wund et al., 2003). The chemical environment of GFP

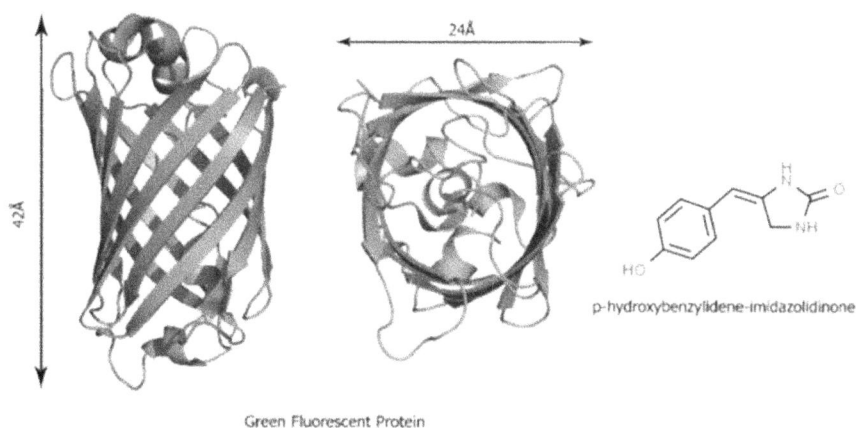

FIGURE 23.1 Structure of green fluorescent protein and the chromophore.

should be maintained for its proper function it was observed that the synthetic chromophore analog of GFP was devoid of fluorescence (Zhang, Gurtu and Kain, 1996). Initially, GFP was expressed in *Caenorhabditis elegans* (*C. elegans*), two neuron receptors, ALMR and PLMR, were tagged with GFP and fluorescence was stable; however, photobleaching occurred. Later it was used in many localization studies mainly including the cell wall synthesis of *E. coli*, Proteins such as FtsA, FtsZ and ZipA had fused to FP, and colocalization of the protein was identified. Similar to the protein DisA whose function was found using the fluorescent protein. In the Gram-positive bacteria *Bacillus subtilis*, a process of delayed cell cycle response to DNA damage was identified with GFP. DisA scan for DNA damage before the onset of sporulation. GFP-tagged DisA protein was found to colocalize with DNA in the sporulation-induced bacteria. The movement of the DisA protein was imaged using time-lapse microscopy (Bejerano-Sagie et al., 2006). GFP becomes a remarkable marker; however, the sensitivity and poor fluorescent intensity of WT-GFP is low. To increase the sensitivity of GFP, codon optimization was performed to produce 35 times brighter than WT-GFP. Which was then called Enhanced GFP (EGFP) (Cormack, Valdivia and Falkow, 1996).

23.3   FLUORESCENT PROTEIN VARIANTS

Properties of fluorescent protein are improved by a mutation which results in improved fluorescence, stability, maturation rate, emission color, and other physicochemical properties. Initially, maturation of WT-GFP occurs at 28°C later optimized to 37°C subsequently reducing the fluorescence and maturation rate. Further mutation of F64L causes a low maturation rate. Some other important mutations that made changes in the properties of GFP are listed in Table 23.2.

23.4   FLUORESCENT PROTEIN – TYPES

Random and site-directed mutations in WT-GFP not only alters the properties of GFP but also change the excitation and emission region of the chromophore resulting in color variants of GFP and FPs from different species such as Discosoma (dsRed) and Anemonia majano (AmCyan1) resulted in the wide color palette. Photostability, sensing property, photo-switch ability and expanded spectrum were color variants of FPs that assist in targeting two or more proteins in a cell. mVenus and mCherry were fused to two different proteins and monitored. Similarly, mCherry, EGFP, mTagBFP, and mTurquoise can be used in combination to track different proteins in a single cell. The excitation and emission spectrum of these mutant proteins differ from each other, using filters the expression of each protein can be focalized individually Table 23.3.

23.5   IMAGING TECHNIQUES

Time-lapse and molecular movement studies were initially performed using widefield microscopy. The simplest and most widely used technique has the major

TABLE 23.2
Mutants of FPs That Show Enhanced Properties

Critical mutation	Native protein	Properties	Reference
F64L	WT-GFP	Improved maturation at 37°C	(Patterson, Knobel, Sharif, Kain & Piston, 1997; Pédelacq, Cabantous, Tran, Terwilliger & Waldo, 2006)
S30R	WT-GFP	Faster folding rate	(Griesbeck, Baird, Campbell, Zacharias & Tsien, 2001)
Q69M	The yellow mutant of GFP	Improves photostability and resistance to chloride and low pH	(Cubitt, Woollenweber & Heim, 1998)
S72A, N149K	WT-GFP	Enhances folding and stability	(Born & Pfeifer, 2019)
S147P	smRS-GFP (a variant of GFP)	Fast maturation	(Crameri, Whitehorn, Tate, & Stemmer, 1996)
V163A	WT-GFP	Reduce hydrophobicity	(Watanabe, 2007)
I167T	WT-GFP	Reduced thermosensitivity	(Remington, 2006)

disadvantage of focus signal. Later fluorescence and epifluorescence microscopes were developed which have their limitation of excitation power (Paddock, 2000) and poor resolution. Later, Confocal Laser Scanning Microscopy (CLSM) with high-resolution optical imaging and depth selectivity. A key feature of CLSM is "optical sectioning", the ability to acquire images at different depths within the sample. Which is an advantage compared to other microscopy techniques (Miyawaki et al., 1997; Wright and Wright, 2002). Fluorescent imaging after the 1970s uses energy transfer, photobleaching, and photoactivation-based techniques such as FRET, FRAP, FLIP, and FLAP.

23.5.1 FORSTER RESONANCE ENERGY TRANSFER

It is based on the transfer of energy via long-range dipole–dipole coupling (non-radioactively) from a donor to the acceptor. Distance between the acceptor and donor should be less than or approximately equal to 10nm. The resolution limit of the FRET was higher than other optical microscopy. When the emission spectrum and absorption spectrum of the donor and acceptor overlap, fluorescent resonance energy transfer takes place (Figure 23.2). The dipole of the donor and acceptor fluorophore is in mutual orientation because of the energy transfer constraint by distance. FRET uses pairs of chromophores for the studies which are commonly called "FRET pairs" or "FRET couples".

TABLE 23.3

Different Color Variants of FPs and Their Excitation and Emission Spectra

Color variants	Molecular weight (kDa)	Excitation spectra	Emission spectra	PDB ID	Organism	References
mCherry	26.72	587	610	2H5Q	*Discosoma sp.*	(Arpino, Rizkallah & Jones, 2012)
mEGFP	27	488	507	4EUL	*Aequorea victoria*	(Kremers, Goedhart, van Munster & Gadella, 2006)
mVenus	26.9	515	528	1EMA	*Aequorea victoria*	(Subach et al., 2008)
mTagBFP	26.7	399	456	3M24	*Entacmaeaquadricolor*	(Shcherbo et al., 2007)
mKate2	26.07	588	633	3BXA	*Entacmaeaquadricolor*	(Shaner et al., 2008)
mApple	27	568	592	1ZGO	*Discosoma sp.*	(Goedhart et al., 2010)
mTurquoise	26.9	434	474	4AR7	*Aequorea victoria*	(Shaner et al., 2013)
mNeonGreen	26.65	506	517	5LTR	*Branchiostomalanceolatum*	(Schawlow & Townes, 1958)

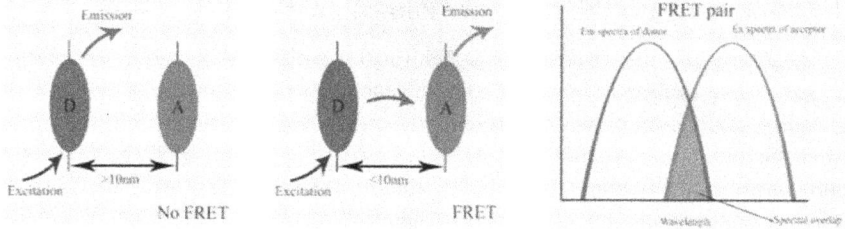

FIGURE 23.2 The emission energy of the donor molecule (D) is used as the excitation energy of the acceptor (A). This happens only when the emission spectra of the donor and the absorption spectra of the acceptor overlap and the distance between the molecule are less than 10nm.

The first effective fluorescent proteins that are used as FRET pair are cyan fluorescent protein and yellow fluorescent protein as donor and acceptor respectively. Calmodulin was tagged to CYF and the calmodulin-binding peptide M13 was tagged to the YFP when Ca^{2+} binds, calmodulin wrapped around M13 subsequently enhances the transfer of energy (FRET) between the FPs Table 23.4 (Mueller, Morisaki, Mazza and McNally, 2012).

23.5.2 CHALLENGES

Information on the distance between the molecules is not predicted using FRET because it is limited to a distance of ~10nm. Low and fluctuating signals and photobleaching limit the use of FRET. A major challenge in FRET imaging is the imaging of multiple fluorescent protein-based FRET couples in the same cell to obtain multiple parameter information concomitantly.

TABLE 23.4
Commonly Used FRET Pairs

FRET pairs		Excitation spectra	Emission spectra	Forster distance
Donor	Acceptor	(nm)	(nm)	(nm)
BFP	DsRFP	380	586	3.1–3.3
CFP	GFP	433	526	4.7–4.9
EBFP2	mEGFP	383	507	4.8
CFP	YFP	433	526	5.0
GFP	YFP	475	526	5.5–5.7
GFP	mRFP	475	579	4.7
TFP1	mVenus	492	528	5.1
EGFP	mCherry	507	510	5.1
Venus	tdTomato	528	581	5.9
Venus	mCherry	528	610	5.7
Venus	mPlum	528	649	5.2

23.5.3 FLUORESCENT RECOVERY AFTER PHOTOBLEACHING

In FRAP, a small area of the cell is illuminated with high light intensity with a focused laser beam and irreversible photobleaching takes place in the fluorescent molecules. This is then later recovered by the non-bleached fluorescent molecules in that surrounding at a particular velocity. During the photobleaching of chromophores, irreversible damage to the fluorescent molecule permanently interrupts the cycle of excitation and emission. This is therefore recovery of the fluorophore possible only by the non-photobleached fluorescence through the process of diffusion between them (Sekar and Periasamy, 2003).

23.5.3.1 Limitation of FRAP

Photo switching was one of the common limitations to the use of FRAP, it is a form of reversible photobleaching where fluorescence recovery takes place by itself after some time instead of recovery by the unbleached fluorescent molecules. Photo switching was reported in GFP and its variants (Yuan and Axelrod, 1994). The possibility of heating by the laser with FRAP was found to be another concern. For the bleaching of the fluorophore, high-power light sources were used leading to localized heating that can be reduced by focusing a laser for a short time, normally 20msec or less (Ishikawa-Ankerhold, Ankerhold and Drummen, 2012).

23.6 FLUORESCENT LOSS IN PHOTOBLEACHING

A complementary method to FRAP, the fluorescent molecule is repeatedly photobleached at a small region to prevent fluorescence recovery. Loss in fluorescence in that region elucidates the mobility of the fluorescently-tagged protein. It is highly useful in the study of the movement of molecules across the cellular compartment. Active and passive transport of molecules across the membrane can be studied with the combination of FRAP and FLIP. A major disadvantage of the photobleaching-based method is that it is not possible to trace all labeled molecules. The bleached molecules cannot be visualized consequently (Dunn, Dobbie, Monypenny, Holt and Zicha, 2002).

23.7 FLUORESCENT LOCALIZATION AFTER PHOTOBLEACHING

It requires two fluorescent molecules, one is locally photobleached and another is used as a reference both are imaged independently or simultaneously with fluorescent microscopy. Absolute FLAP signaling can be obtained by subtracting the signals from bleached molecules allowing the tracking of the label molecule (van Royen et al., 2007). These methods are also used in combination and variations are made to enhance the study of molecules. For example, FRET and FRAP were used in combination to yield a technique that investigates the mobility of interacting molecules (Chen and Huang, 2001; Royen, Dinant, Farla, Trapman and Houtsmuller, 2009). FRET and FLIP were used in combination to determine the mobility of GFP-tagged protein (Reid and Flynn, 1997).

23.7.1 LIMITATIONS OF USING TAGS

It requires cDNA or a clone. The structure or function of the protein may be disturbed by the epitope tags. Expression of the epitope-tagged protein may be at abnormal levels because of heterologous promoters. GFP cannot be used in anaerobic bacteria as the maturation of the chromophore requires O^2 (Inouye and Tsuji, 1994; Reid and Flynn, 1997). Both the POI and GFP must be functional when expressed. Overproduction of the GFP-fused proteins results in localization at physiologically irrelevant locations and localizes to the part of the cell that does not localize when normally synthesized. Complimentary studies with GFP-tagged proteins should be treated with caution because the properties of the native protein were altered after fusion which may not be predicted.

23.8 MAJOR CHALLENGES IN FPS

Several criteria play a major role in localization studies mainly the copy number of genes and the duration of expression of the tagged protein. The transcriptional promoter and enhancer strength in the expression of the gene. It is advised to optimize the codon of the FPs to that of the host. Hindrance to the folding of the fusion protein along with the POI is limited although superfold GFP is available. Dimerization, photobleaching, photoisomerization, autofluorescence of the cell, noise and background signals are the other factors that limit the use of FPs in localization studies.

23.9 CONCLUSION

Advances in microscopic imaging depict the perception of the cellular structures and dynamics happening in live cells and may give an understanding of *in vivo* complex biological processes. Several approaches such as a chemical biology-based approach have been developed for several needs for fluorescent labeling of POIs in live cells. Small-sized tags produce efficient functions of POI while larger tags may lead to damage to the functions of POI. These labeling reactions have proceeded with fast kinetics for the betterment of resolution in biological assays.

Biology combined with imaging and spectroscopy serves as a powerful tool for clinical application, screening of pathogens and research (Govindarajan, Viswalingam, Meganathan and Kandaswamy, 2020; Govindarajan and Kandaswamy, 2022; Govindarajan et al., 2022; Shanmugasundarasamt et al., 2022). FRET, FRAP, FLIM techniques, and fluorescent reagents such as GFP, which are widely used for the activity of genes, metabolic or signaling pathways, and cellular components, became versatile tools for imaging. On the other hand, self-labeling enzyme tags provide high specificity, small size, and covalent labeling with its limitations such as the requirement of high substrate concentration and incubation time. The covalent labeling technique is fast yet limited to labeling the N-terminal of cell surface proteins. This chapter will guide the researchers in conducting studies in live-cell functioning such as cellular events and protein behavior.

REFERENCES

Arpino, J. A., Rizkallah, P. J., & Jones, D. D. (2012). Crystal structure of enhanced green fluorescent protein to 1.35 Å resolution reveals alternative conformations for Glu222, e47132.

Bacart, J., Corbel, C., Jockers, R., Bach, S., & Couturier, C. (2008). The BRET technology and its application to screening assays. *Biotechnology Journal: Healthcare Nutrition Technology*, 3(3), 311–324.

Bejerano-Sagie, M., Oppenheimer-Shaanan, Y., Berlatzky, I., Rouvinski, A., Meyerovich, M., & Ben-Yehuda, S. (2006). A checkpoint protein that scans the chromosome for damage at the start of sporulation in Bacillus subtilis. *Cell*, 125(4), 679–690.

Bendezú, F. O., Hale, C. A., Bernhardt, T. G., & De Boer, P. A. (2009). RodZ (YfgA) is required for proper assembly of the MreB actin cytoskeleton and cell shape in E. coli. *The EMBO Journal*, 28(3), 193–204.

Berk, V., Fong, J. C., Dempsey, G. T., Develioglu, O. N., Zhuang, X., Liphardt, J., ... Chu, S. (2012). Molecular architecture and assembly principles of Vibrio cholerae biofilms. *Science*, 337(6091), 236–239.

Bisson-Filho, A. W., Hsu, Y.-P., Squyres, G. R., Kuru, E., Wu, F., Jukes, C., ... VanNieuwenhze, M. S. (2017). Treadmilling by FtsZ filaments drives peptidoglycan synthesis and bacterial cell division. *Science*, 355(6326), 739–743.

Born, J., & Pfeifer, F. (2019). Improved GFP variants to study gene expression in haloarchaea. *Frontiers in Microbiology*, 10, 1200.

Chen, D., & Huang, S. (2001). Nucleolar components involved in ribosome biogenesis cycle between the nucleolus and nucleoplasm in interphase cells. *The Journal of Cell Biology*, 153(1), 169–176.

Coons, A. H., Creech, H. J., & Jones, R. N. (1941). Immunological properties of an antibody containing a fluorescent group. *Proceedings of the Society for Experimental Biology and Medicine*, 47(2), 200–202.

Cormack, B. P., Valdivia, R. H., & Falkow, S. (1996). FACS-optimized mutants of the green fluorescent protein (GFP). *Gene*, 173(1), 33–38.

Crameri, A., Whitehorn, E. A., Tate, E., & Stemmer, W. P. (1996). Improved green fluorescent protein by molecular evolution using DNA shuffling. *Nature Biotechnology*, 14(3), 315–319.

Cubitt, A. B., Woollenweber, L. A., & Heim, R. (1998). Understanding structure—Function relationships in the Aequorea victoria green fluorescent protein. *Methods in Cell Biology*, 58, 19–30.

Doi, N., Takashima, H., Kinjo, M., Sakata, K., Kawahashi, Y., Oishi, Y., ... Endo, Y. (2002). Novel fluorescence labeling and high-throughput assay technologies for in vitro analysis of protein interactions. *Genome Research*, 12(3), 487–492.

Dragulescu-Andrasi, A., Chan, C. T., De, A., Massoud, T. F., & Gambhir, S. S. (2011). Bioluminescence resonance energy transfer (BRET) imaging of protein–protein interactions within deep tissues of living subjects. *Proceedings of the National Academy of Sciences of the United States of America*, 108(29), 12060–12065.

Dunn, G., Dobbie, I., Monypenny, J., Holt, M., & Zicha, D. (2002). Fluorescence localization after photobleaching (FLAP): A new method for studying protein dynamics in living cells. *Journal of Microscopy*, 205(1), 109–112.

Fairclough, R. H., & Cantor, C. R. (1978). The use of singlet-singlet energy transfer to study macromolecular assemblies. *Methods in Enzymology*, 48, 347–379.

Follenius-Wund, A., Bourotte, M., Schmitt, M., Iyice, F., Lami, H., Bourguignon, J.-J., ... Pigault, C. (2003). Fluorescent derivatives of the GFP chromophore give a new insight into the GFP fluorescence process. *Biophysical Journal*, 85(3), 1839–1850.

Gautier, A., Juillerat, A., Heinis, C., Corrêa Jr, I. R., Kindermann, M., Beaufils, F., & Johnsson, K. (2008). An engineered protein tag for multiprotein labeling in living cells. *Chemistry and Biology*, *15*(2), 128–136.

Goedhart, J., Van Weeren, L., Hink, M. A., Vischer, N. O., Jalink, K., & Gadella, T. W. (2010). Bright cyan fluorescent protein variants identified by fluorescence lifetime screening. *Nature Methods*, *7*(2), 137–139.

Govindarajan, D. K., & Kandaswamy, K. (2022). Virulence factors of uropathogens and their role in host pathogen interactions. *The Cell Surface*, 8, 100075.

Govindarajan, D. K., Meghanathan, Y., Sivaramakrishnan, M., Kothandan, R., Muthusamy, A., Seviour, T. W., & Kandaswamy, K. (2022). *Enterococcus faecalis* thrives in dual-species biofilm models under iron-rich conditions. *Archives of Microbiology*, 204(12), 710.

Govindarajan, D. K., Viswalingam, N., Meganathan, Y., & Kandaswamy, K. (2020). Adherence patterns of Escherichia coli in the intestine and its role in pathogenesis. *Medicine in Microecology*, 5, 100025.

Griesbeck, O., Baird, G. S., Campbell, R. E., Zacharias, D. A., & Tsien, R. Y. (2001). Reducing the environmental sensitivity of yellow fluorescent protein: Mechanism and applications. *Journal of Biological Chemistry*, *276*(31), 29188–29194.

Inouye, S., & Tsuji, F. I. (1994). Aequorea green fluorescent protein: Expression of the gene and fluorescence characteristics of the recombinant protein. *FEBS Letters*, *341*(2–3), 277–280.

Ishikawa-Ankerhold, H. C., Ankerhold, R., & Drummen, G. P. (2012). Advanced fluorescence microscopy techniques—Frap, Flip, Flap, Fret and flim. *Molecules*, *17*(4), 4047–4132.

Janssen, D. B. (2004). Evolving haloalkane dehalogenases. *Current Opinion in Chemical Biology*, *8*(2), 150–159.

Kandaswamy, K., Liew, T. H., Wang, C. Y., Huston-Warren, E., Meyer-Hoffert, U., Hultenby, K., … Henriques-Normark, B. (2013). Focal targeting by human β-defensin 2 disrupts localized virulence factor assembly sites in Enterococcus faecalis. *Proceedings of the National Academy of Sciences of the United States of America*, *110*(50), 20230–20235.

Ke, N., Landgraf, D., Paulsson, J., & Berkmen, M. (2016). Visualization of periplasmic and cytoplasmic proteins with a self-labeling protein tag. *Journal of Bacteriology*, *198*(7), 1035–1043.

Kline, K. A., Kau, A. L., Chen, S. L., Lim, A., Pinkner, J. S., Rosch, J., … Beatty, W. (2009). Mechanism for sortase localization and the role of sortase localization in efficient pilus assembly in Enterococcus faecalis. *Journal of Bacteriology*, *191*(10), 3237–3247.

Kremers, G.-J., Goedhart, J., van Munster, E. B., & Gadella, T. W. (2006). Cyan and yellow super fluorescent proteins with improved brightness, protein folding, and FRET Förster radius. *Biochemistry*, *45*(21), 6570–6580.

Liss, V., Barlag, B., Nietschke, M., & Hensel, M. (2015). Self-labelling enzymes as universal tags for fluorescence microscopy, super-resolution microscopy and electron microscopy. *Scientific Reports*, *5*(1), 1–13.

Los, G. V., Encell, L. P., McDougall, M. G., Hartzell, D. D., Karassina, N., Zimprich, C., … Urh, M. (2008). HaloTag: A novel protein labeling technology for cell imaging and protein analysis. *ACS Chemical Biology*, *3*(6), 373–382.

Lotze, J., Reinhardt, U., Seitz, O., & Beck-Sickinger, A. G. (2016). Peptide-tags for site-specific protein labelling in vitro and in vivo. *Molecular Biosystems*, *12*(6), 1731–1745.

Maurel, D., Comps-Agrar, L., Brock, C., Rives, M.-L., Bourrier, E., Ayoub, M. A., … Prézeau, L. (2008). Cell-surface protein-protein interaction analysis with time-resolved FRET and snap-tag technologies: Application to GPCR oligomerization. *Nature Methods*, *5*(6), 561–567.

Miyawaki, A., Llopis, J., Heim, R., McCaffery, J. M., Adams, J. A., Ikura, M., & Tsien, R. Y. (1997). Fluorescent indicators for Ca2+ based on green fluorescent proteins and calmodulin. *Nature*, *388*(6645), 882–887.

Mueller, F., Morisaki, T., Mazza, D., & McNally, J. G. (2012). Minimizing the impact of photoswitching of fluorescent proteins on FRAP analysis. *Biophysical Journal*, *102*(7), 1656–1665.

Munro, S., & Pelham, H. (1984). Use of peptide tagging to detect proteins expressed from cloned genes: Deletion mapping functional domains of Drosophila hsp 70. *The EMBO Journal*, *3*(13), 3087–3093.

Ormö, M., Cubitt, A. B., Kallio, K., Gross, L. A., Tsien, R. Y., & Remington, S. J. (1996). Crystal structure of the Aequorea victoria green fluorescent protein. *Science*, *273*(5280), 1392–1395.

Paddock, S. W. (2000). Principles and practices of laser scanning confocal microscopy. *Molecular Biotechnology*, *16*(2), 127–149.

Patterson, G. H., Knobel, S. M., Sharif, W. D., Kain, S. R., & Piston, D. W. (1997). Use of the green fluorescent protein and its mutants in quantitative fluorescence microscopy. *Biophysical Journal*, *73*(5), 2782–2790.

Paulmurugan, R., Umezawa, Y., & Gambhir, S. (2002). Noninvasive imaging of protein–protein interactions in living subjects by using reporter protein complementation and reconstitution strategies. *Proceedings of the National Academy of Sciences of the United States of America*, *99*(24), 15608–15613.

Pédelacq, J.-D., Cabantous, S., Tran, T., Terwilliger, T. C., & Waldo, G. S. (2006). Engineering and characterization of a superfolder green fluorescent protein. *Nature Biotechnology*, *24*(1), 79–88.

Perkovic, M., Kunz, M., Endesfelder, U., Bunse, S., Wigge, C., Yu, Z., ... Malkusch, S. (2014). Correlative light-and electron microscopy with chemical tags. *Journal of Structural Biology*, *186*(2), 205–213.

Reid, B. G., & Flynn, G. C. (1997). Chromophore formation in green fluorescent protein. *Biochemistry*, *36*(22), 6786–6791.

Remington, S. J. (2006). Fluorescent proteins: Maturation, photochemistry and photophysics. *Current Opinion in Structural Biology*, *16*(6), 714–721.

Royen, M. E. V., Dinant, C., Farla, P., Trapman, J., & Houtsmuller, A. B. (2009). FRAP and FRET methods to study nuclear receptors in living cells. In *The Nuclear Receptor Superfamily* (Ed. J. Lain) (pp. 69–96). Springer. McEwan, New Jersey.

Schawlow, A. L., & Townes, C. H. (1958). Infrared and optical masers. *Physical Review*, *112*(6), 1940.

Schermelleh, L., Heintzmann, R., & Leonhardt, H. (2010). A guide to super-resolution fluorescence microscopy. *Journal of Cell Biology*, *190*(2), 165–175.

Sekar, R. B., & Periasamy, A. (2003). Fluorescence resonance energy transfer (FRET) microscopy imaging of live cell protein localizations. *The Journal of Cell Biology*, *160*(5), 629.

Shaner, N. C., Lambert, G. G., Chammas, A., Ni, Y., Cranfill, P. J., Baird, M. A., ... Israelsson, M. (2013). A bright monomeric green fluorescent protein derived from Branchiostoma lanceolatum. *Nature Methods*, *10*(5), 407–409.

Shaner, N. C., Lin, M. Z., McKeown, M. R., Steinbach, P. A., Hazelwood, K. L., Davidson, M. W., & Tsien, R. Y. (2008). Improving the photostability of bright monomeric orange and red fluorescent proteins. *Nature Methods*, *5*(6), 545–551.

Shanmugasundarasamy, T., Govindarajan, D. K., & Kandaswamy, K. (2022). A review on pilus assembly mechanisms in Gram-positive and Gram-negative bacteria. *The Cell Surface*, 8, 100077.

Shcherbo, D., Merzlyak, E. M., Chepurnykh, T. V., Fradkov, A. F., Ermakova, G. V., Solovieva, E. A., ... Lukyanov, S. (2007). Bright far-red fluorescent protein for whole-body imaging. *Nature Methods, 4*(9), 741–746.

Shimomura, O., Johnson, F. H., & Saiga, Y. (1962). Extraction, purification and properties of aequorin, a bioluminescent protein from the luminous hydromedusan, Aequorea. *Journal of Cellular and Comparative Physiology, 59*(3), 223–239.

Subach, O. M., Gundorov, I. S., Yoshimura, M., Subach, F. V., Zhang, J., Grüenwald, D., ... Verkhusha, V. V. (2008). Conversion of red fluorescent protein into a bright blue probe. *Chemistry and Biology, 15*(10), 1116–1124.

van Royen, M. E., Cunha, S. M., Brink, M. C., Mattern, K. A., Nigg, A. L., Dubbink, H. J., ... Houtsmuller, A. B. (2007). Compartmentalization of androgen receptor protein–protein interactions in living cells. *The Journal of Cell Biology, 177*(1), 63–72.

Watanabe, S. (2007). Fluorescent indicators for Ca2+ based on green fluorescent proteins and calmodulin. *Tanpakushitsu Kakusan Koso. Protein, Nucleic Acid, Enzyme, 52*(13 Suppl), 1770–1771.

Wright, S. J., & Wright, D. J. (2002). Introduction to confocal microscopy. *Cell Biological Applications of Confocal Microscopy, in Methods in Cell Biology, 70,* 1–85.

Yuan, Y., & Axelrod, D. (1994). Photobleaching with a subnanosecond laser flash. *Journal of Fluorescence, 4*(2), 141–151.

Zhang, G., Gurtu, V., & Kain, S. R. (1996). An enhanced green fluorescent protein allows sensitive detection of gene transfer in mammalian cells. *Biochemical and Biophysical Research Communications, 227*(3), 707–711.

Index

For Product Safety Concerns and Information please contact our EU
representative GPSR@taylorandfrancis.com
Taylor & Francis Verlag GmbH, Kaufingerstraße 24, 80331 München, Germany

www.ingramcontent.com/pod-product-compliance
Lightning Source LLC
Chambersburg PA
CBHW060348220326
41598CB00023B/2848

9 7 8 1 0 3 2 3 1 0 3 1 2